Eliza Barton Lyman

The Coming Woman or the Royal Road to Physical Perfection

A Series of Medical Lectures

Eliza Barton Lyman

The Coming Woman or the Royal Road to Physical Perfection
A Series of Medical Lectures

ISBN/EAN: 9783337270094

Printed in Europe, USA, Canada, Australia, Japan

Cover: Foto ©berggeist007 / pixelio.de

More available books at **www.hansebooks.com**

E. B. Lyman,

OR,

THE ROYAL ROAD TO PHYSICAL PERFECTION.

A SERIES OF MEDICAL LECTURES

BY

ELIZA BARTON LYMAN,

LECTURER AND TEACHER OF ANATOMY,
PHYSIOLOGY AND HYGIENE.

LANSING, MICH.:
W. S. GEORGE & CO., PRINTERS AND BINDERS.
1880.

To

𝔐rs. 𝔍ames 𝔖. 𝔓hillips,

the

Friend of Many Winters,

I,

With Love and Respect,

Dedicate

This Little Volume.

"She hath wrought this in love of her kind."

CAROLINE PHILLIPS.

THE COMING WOMAN.

How you may know her. She shall possess physical and mental equipoise; she shall be the co-laborer with and equal of man; she shall be the patron of Art, Science, and Literature; she shall unite factions, cement home interests; shall be the friend and guide of the youth; she shall love, reverence, defend, and advance the interests of all women.

She shall silence the tongue of slander, plead the cause of the oppressed, seek out and sustain the fallen; she shall condemn no man or woman; she shall be strong, and wise, and restful. Her speech shall be golden, and her crown love.

She will not come along your highways, heralded by the clang of trumpets, but up through your by-ways, and beneath her footsteps shall spring violets, daisies, and clover bloom, and all sweet things that breathe of Peace, Purity, Justice, Mercy, and Truth.

PREFACE.

THIS work is especially addressed to women, not simply because they most need it,—neither because of the Freemasonry existing between souls linked together by the common bonds of sex bearing like crosses and wrongs; but from a feeling of grateful appreciation and love. For the noblest ambitions of my life, my strongest incentives to greater excellence in every direction, have been given by women,—many times in acts insignificant in themselves, but expressing much to me. The kindly pressure of the hands, the eye messages of interest and sympathy, little words of encouragement and affection, a simple nosegay of fragrant flowers, courtesies forgotten doubtless as soon as performed, but whose influence still remains, having been my stimulus to this effort.

And now, to the world of womankind everywhere, I extend this, my right hand of fellowship and good faith.

E. B. L.

INTRODUCTION.

THERE are few subjects of greater importance to mankind than the study of the structure, forms, and uses of the various portions of the human body. Not only is it important, but intensely interesting, as the crowning work of the Creator's hand. The human body is the earthly home of the immortal soul, and perfectly arranged and adapted to meet its manifold needs. This body is subject to all of the changes incident to matter,—that of growth, development, and decay. It is controlled by fixed laws similar to those governing all of the forms in the subordinate kingdoms. The science of human physiology, like all other natural sciences, can be readily comprehended by all classes when presented in a simple, plain manner, unencumbered with technicalities. This study is not more abstruse than those pertaining to every-day life with which men and women familiarize themselves. All persons should at least possess an outline knowledge of Physiology and Hygiene, sufficient to enable them to guard against accident and disease.

That physicians alone should monopolize this branch of knowledge is manifestly absurd, as much so as that a limited number of men should possess and hold all of the secrets of the science of Agriculture, Horticulture, or Household Chemistry. We would regard it as ridiculous in the extreme if all of our farmers were compelled to call in the aid of some one man to tell them when to enrich their soil, how to cure their crops, and guard against the weevil, rust, and blight. A knowledge of the physical structure is just as important to, and would be as readily understood by every man, woman, and child, as is the science of cultivating a field of wheat, or a knowledge of the chemical constituents entering into an ordinary meal, or the law controlling the flying of a kite. The study of respiration, digestion, circulation, brain, and nerve action, growth and development of the muscular system, together with the importance of appropriate food, fresh air, and judicious exercise, should be made just as familiar to every child as are the rudiments of the English language or mathematics.

If less Greek and Latin, and more Anatomy, Physiology, and Hygiene were introduced into the curriculum of the young men and women in our institutions of learning, they would be better prepared to solve the problem of life as fathers and mothers of the race. It is a startling fact, that in this country one-fourth of all the children born, die before they reach the age of one year, and one-half before they reach the age of ten. This shocking mortality is the result of ignorance on the part of the parents, and could be obviated by a better knowledge of the laws of health.

In the compilation of this work, the author has sought to simplify the subject as much as possible, and at the same time not detract from its strength. The aim has been to instruct and entertain, without wearying the reader. If it contains but a few seeds of truth, which shall drop into good soil and bring forth fruit for the good of humanity, the work has not been in vain.

TABLE OF CONTENTS.

———◆———

CHAPTER I.

MUSCULAR AND OSSEOUS SYSTEMS.

HYGIENE OF THE MUSCLES.

CHAPTER II.

SKIN AND GLANDULAR SYSTEM.

HYGIENE OF THE SKIN.

CHAPTER III.

CIRCULATION OF THE BLOOD.

CHAPTER IV.

RESPIRATION.

CHAPTER V.

DIGESTION.

CHEMISTRY OF FOOD.

DIETETICS FOR CHILDREN.

EAT, DRINK, AND BE MERRY.

CHAPTER VI.

NERVOUS SYSTEM.

CHAPTER VII.

FEMALE GENERATIVE ORGANS.

CHAPTER VIII.

MATERNITY.

CARE OF PREGNANT WOMEN AND THEIR TREATMENT AFTER CHILDBIRTH.

CARE OF NEW-BORN INFANTS.

CHAPTER IX.

HEREDITY.

CHAPTER X.

MOTHER'S INFLUENCE.

CHAPTER XI.

WILL POWER.

CHAPTER XII.

UNSEEN INFLUENCES.

CHAPTER XIII.

PHYSICAL PERFECTION.

CHAPTER XIV.

THE TOILET.

THE SCALP.

THE TEETH.

CARE OF THE EYE.

THE EAR.

CHAPTER XV.

DISEASES AND THEIR REMEDIAL AGENTS—HOME TREATMENT FOR COLDS, SORE THROAT, AND VARIOUS DISORDERS.

SCARLET FEVER.

How it should be treated—Temperature of the room—Sunlight—Disinfectants—Hot packs—Diet—Palatable and cooling drinks—Convalescing.

HEADACHES.

Magnetic treatment—Counter irritants—Hot and cold applications—Predisposing causes—How removed—Sedentary habits.

RHEUMATISM AND ITS TREATMENT.

Rheumatic tendencies hereditary—How they are developed—Causes of rheumatism—Bone and joint rheumatism—Bathing—Sunbaths—Flannel garments—Use of acids—Simple and often unfailing remedies—Damp and shaded houses—Effect of heat and cold.

FAINTING.

Cause—Treatment—How it may be avoided—Influence of the mind—Exercises that are beneficial to both mind and body.

HYSTERIA.

Causes—Remedies—False ideas—Common sense view.

CANCER.

How it may be cured without the use of the knife—Hereditary predisposition—How this may be overcome—Simple remedies.

SCROFULA, ERYSIPELAS, AND SALT RHEUM.

Causes and cures—Hygienic treatment.

DYSENTERY AND DIARRHŒA.

Treatment—Chronic diarrhœa—Aggravating causes.

·

THE COMING WOMAN.

CHAPTER I.

MUSCULAR SYSTEM.

THE framework of the body is composed of bones, something over two hundred in number, so arranged as to sustain the weight of the various internal organs and protect them from injury. The more important and delicate the organ,‘ the more carefully it is shielded. For example, the brain is enclosed within a strong bony box; the spinal marrow is surrounded and guarded by its twenty-four strongly spined bony rings; the eye and ear are also protected with thick bones; the heart and lungs have a flexible coat of mail covering them in form of breast bone, ribs, and shoulder blades—this group of bones being so arranged that they can be easily moved in the act of respiration. This bony foundation is connected and held together by firm bands of muscles, and strong white glistening cords known as tendons. One portion of the muscles is entirely under the

control of the will, while another acts independently.
These classes are denominated voluntary and involuntary, which not only perform the leverage of the body,
but serve also as outside walls and internal partitions,
and give permanence to the entire structure.

There are about five hundred of these beautifully
constructed elastic fibres, all splendidly adapted, and
we believe wisely intended for use. They are constructed in a spiral form, which enables them to contract and expand easily, and in order to keep them
flexible and in health they must be given a reasonable
amount of exercise daily, so that the life of the blood
may be formed into muscular structures, instead of
being converted into fatty deposits, which if predomi-
nating, impedes all of the bodily functions, as an excess
of fat is as much an indication of disease as an undue
emaciation would be. Much of the prevailing heart
difficulty is the result of the muscular embarrassment
growing out of an overload of fat surrounding that
organ, just as an excess of adipose on the limbs and
body impedes locomotion or vigorous exercise of any
kind.

Dr. Winship was in the habit of determining the
health of an individual by the tone of the muscles
alone, claiming that the health deteriorated in proportion as they lost their firmness and force.

Each person is endowed with a like number of these flexible fibres, and the difference between the puny stripling and the gymnast is only a matter of development. In the former the muscular fibres are weak, flabby and bloodless; in the latter, firm, well coiled up, and charged with nutrient blood, and ready for all the demands which the soul may make on the body. The female structure has the same number of muscles as that of the male, constructed upon the same plan and intended for a similar use, which failing to receive, become weak, and inadequate for the work designed for them. There is no good reason why women should have feeble, useless muscles any more than that men should. The physical perfection and beauty of the ancient Greek women have served as models for the painter and sculptor for over two thousand years. It is said that Lycurgus ordered the women of Sparta to be sent to the gymnasium, and slaves only to be put to the embroidery frame and spinning-wheel; and it is a historical fact that the women of Greece exercised as regularly in the gymnasiums as did the men, and doubtless the strength, symmetry, and beauty of the men were due even more to the physical vigor of the mother than to any training they may have received in their youth.

There is one point which we particularly desire to impress upon the mind of the reader, and that is the

5

importance of regular and systematic exercise. The bodily functions are largely carried on by the muscular structures. Not only are the sweat tubes and absorbents assisted by them in their work, but the blood veins also receive a new impulse, which enables them to lift their load of impure blood inward and upward towards the heart and lungs.

The glandular system depends greatly upon the muscular for assistance in the performance of its functions. For example, the liver is kept in an active condition by the rolling motion of the stomach underneath and the contraction and expansion of the diaphragm above. Under the arm there is a large group of glands, whose office is to expel from the body a disagreeable and poisonous refuse matter. Every form of arm exercise aids this group in the performance of its duty by the pressure of the muscles on the glands. In the lower part of the abdomen and inside of the thighs there are also great numbers of small glands having a similar office to perform, which the exercise of walking greatly facilitates by the contraction and expansion of the walking muscles. All body movements assist the kidneys in expelling their secretions, and also in dislodging any earthy matter from the tubules which convey the urine from the body of the kidneys into the ducts leading to the bladder. Nature evidently designed that this work should be accomplished by the expansion and contraction of the muscles in the healthy performance of their various offices.

HOW TO RESTORE **THE** MUSCLES.

All persons desiring to increase their physical stamina **should** make a *thorough* study and practice of Elocution **in all of its branches.** Not in the sense that it is usually accepted, merely for **the purpose of** public reading, but **as a means of** physical and mental development. When properly taught it embraces all of **the forms of** physical training necessary for health **and** symmetry. In it **we** find a full range of voice culture, instructions in deep breathing, which tends to enlarge the lung capacity and give tone to the muscular **tissues.** It **performs** all of these missions to the body, **to** say nothing of the stimulus given to the brain by the contact with the thoughts of great **minds.** The benefits are far greater **than those** derived **from the** study and practice of music alone. Reading aloud **for two** hours each day, **if the** voice is correctly managed, will so strengthen **the vocal** organs that throat difficulties will in most cases disappear.

Owing to the very imperfect arrangement of dress adopted by our country-women, not more than one **in ten** has any control over the breathing muscles situated in the abdomen; depending entirely upon those in the thoracic region, **and** in consequence the breathing is light **and** imperfect; **there is** a loss **of** tonicity in the **heavy respiratory bands which** tends **to** greatly **retard** the action **of all the** organs in the lower viscera, as well as throwing **the strain** upon the thoracic muscles, in which case, **if speaking and** singing **were** frequently indulged in, would seriously impair the vocal cords. Every woman under **fifty** years of age may, by careful **treatment, restore** these **muscles.** A series of voluntary contractions of the abdominal walls will tone up the relaxed fibres, just as rowing tones up the biceps of the **oarsman.** One day's **work** will not accomplish it. **It** will require *persistence,* and if **the** *will* **is weak, put** that under a course of *mental gymnastics.*

Those persons **who, from lack of desire or want of** opportunity, neglect to take regular daily exercise, lose **a most** potent stimulus **to** the entire system, for nothing can ever fully take the place **of** physical exercise. **The** muscles composing **the** internal organs, such as the heart, intestines and diaphragm, sympathize most per- **fectly with** those on the outside **of** the **body. If they** are weak **and** relaxed, **then** the contractions of the heart, stomach, intestines, and breathing muscles are correspondingly weak. Good, firm muscles signify good digestion, good circulation, good, full respiration, perfect assimilation, and consequently a full supply of material for forming new tissues.

No amount of vigorous in-door exertion can adequately compensate for the loss of exercise in the open air. Walking, when the clothing is properly supported from the shoulders, and thick-soled shoes are worn, is the best mode of exercise; it calls into play a greater number of muscles than any other one form of activity; but in walking the arms should be used freely and the clothing be loose enough to permit a swaying motion of the entire body. The American women would do well to imitate their English sisters in this respect, who consider four, five, six, eight and even ten miles only a nice little distance to walk daily. The Princess Louise appears to think ten miles merely an ordinary stroll.

Long continued standing or sitting is far more fatiguing than walking, from the fact that one set of muscles is kept constantly on a strain, while in walking there is a continual expansion and contracting which is rather restful than otherwise. The English women have one class of healthful recreation which the Americans deprive themselves of—that is, horse-back riding. Few days are so stormy as to prevent the English woman from taking her accustomed gallop of fifteen or twenty miles, and to this exhilarating exercise she is largely indebted for her fine physique; while the American women become physically deteriorated by their sedentary and in-door life. Every woman who has the use of her limbs should take a walk each day, gradually increasing the distance as she gains strength. Ordinary storms should impose no barrier, for one can always wrap and protect the body from the inclemency of the weather, and we require fresh air for oxygenizing the blood just as much on a stormy day as on a sunny one, for in the air of living rooms generally there is very little oxygen. Pure air is as absolute a necessity as good food and untainted water; yet how little of the pure, vitalized air people who constantly live in-doors inhale. Nature has a wonderful power of adaptation, otherwise the human family would have been swept off the face of the earth ere this. The vitality sinks to accommodate itself to its surroundings, just as a garrison of soldiers will manage to live for days on short rations, when the stores are becoming exhausted and no immediate prospects of obtaining new supplies. The men *live* but are physically demoralized; so it is with the occupants of ill-ventilated rooms who neglect to refresh the system daily by full, deep breathing in the open air.

All the exercise necessary for the proper development of the body may be obtained from the use of a few simple contrivances that **every** one **can have** at home **at** little cost—less by **far** than **is** spent for useless toys. **Many of** these may **be made** available **in the** parlor or chamber, though all exercises are far more useful in the open air. **A** small portion of the day thus spent **will** afford agreeable recreation as well as useful **exercise.** **The** Indian club, the wand, the ring, and the dumb-bells answer **ordinary** purposes **very** well. Simple contrivances that may be useful for general exercises, and are especially suitable for persons with *weak spines* or **with** spines that **are** the subject of lateral curvature.

CHAPTER II.

SKIN AND GLANDULAR SYSTEM.

OVER the enclosed framework there is arranged a covering of adipose tissue, familiarly known as fat. This fatty coat fills up the depressions left in the joining of the muscular fibres, and gives smoothness and roundness to the figure as well as furnishing heat to the system. In cases of emergency, fleshy persons can live for days on a limited amount of food without experiencing any great inconvenience or loss of strength, the surplus fat of the system serving as fuel.

Over this layer of padding is a soft cushion known as the skin, composed of sebaceous glands or oil sacs, and perspiratory ducts or pores, minute blood vessels or capillaries, and nerve filaments. At the base of this layer of true skin are the absorbents or lymphatic glands, spreading over the entire structure like a network of strung beads. These minute spongy bodies receive from the blood, lymph and saline matter, which if not expelled promptly would induce serious blood irritation, manifesting itself in various eruptive forms. When we examine the skin through the aid of the microscope, we find it made up of two layers, the outer and inner; the inner is called the true skin; the outer, the epidermis or cuticle, which is a thin, tough, elastic, insensible, and semi-transparent coat, composed of minute plates or scales which overlap each other, somewhat like the scales of a fish, layer upon layer, and which are constantly worn off and replaced by new. This outer covering serves as a protection for the sensitive nervous tissues underneath.

The skin is governed by the law of endosmose and exosmose, the same as the lungs, not only affording an escape for the waste products, but also allowing the ingress of the life-giving elements of the air. M. Paul Bert advances the theory, that what is called nerve force, or power, is received from the electric life of the atmosphere, through the millions of nerve filaments on the surface of the body, somewhat as the blood derives its life from the oxygen contained in the respired air. The plates composing the cuticle are, when unobstructed, self-adjusting, rising in such a manner as to permit the perspiratory matter to readily pass out. They also act as guardians to ward off injurious outside influences, by closing suddenly when the heated surface of the body is exposed to intense cold or dampness, thereby preventing the chilling of the sensitive vascular structures and consequent inflammatory condition of the entire system.

These adjustable plates likewise serve to equalize the temperature of the body during the varying changes to which the individual may be subjected. These myriads of windows open and close to suit the requirements of the system; but the hinges need oiling, and the panes washing and polishing, in order that they may serve well the purpose for which they were intended. They must not be permitted to become glued down by the accumulated debris of waste tissues. The important office of the skin has been demonstrated from time to time by experiments upon various animals, by covering the entire body with some impervious coat; death usually ensuing in a few hours after the application.

It is related that at the coronation of one of the popes a little child was chosen to enact the part of an angel, and in order that it should present a dazzling appearance, it was covered from head to foot with gold-foil. The child was soon attacked with nausea and difficulty of breathing, and, although every means was resorted to to restore it, except removing the coating, it died in a short time.

The great suffering experienced by the sudden closing of the pores, on "taking cold," and in attacks of fever, is familiar to all.

The greater portion of the skin's product is water, which passes out through the countless tiny sweat glands and tubes, each of which, if uncoiled, would measure from one-eighth to one-tenth of an inch in length. They are so numerous that if they were placed in a continuous line they would form several miles of tubing, through which a considerable amount of solid waste matter, such as oil, lime, salt, acid, etc., as well as water and the various gases, are being constantly expelled.

This thorough system of drainage may be better appreciated when we take into consideration the fact that the larger portion of the body is composed of water, fully five-eighths, at the least calculation. Considerably over one-half of all the bread, meat and vegetables used as food is simply water, to say nothing of what is consumed in the way of drinks.

At a recent meeting of the "American Medical Association" the following item was given in one report: "Of every seven pounds of food and drink taken into the stomach, five out of the seven passes out through the pores of the skin as waste." From a healthy skin will be eliminated about three pints of water in twenty-four hours, in form of insensible perspiration, and during exercise or exposure to heat a still greater quantity will be lost in form of sensible perspiration.

This continual taking on and throwing off is an imperative necessity of the organism, from the fact that the supply of fresh, pure water taken into the system to-day will by to-morrow have become impure and lifeless, and must needs be replaced by a new installment. Water in the tissues of the body serves as a highway for the conveyance of more important materials, essences, and subtile fluids, among the number electricity, it having an especial affinity for water, which having served its purpose as conductor, is thrown out of the system through the pores of the skin, the kidneys, the mucous linings, and in exhalations from the lungs; it, obeying the law of attraction, makes its exit from the body through every possible avenue of escape to seek and mingle with the great world of water without. The entire structure is so porous that the fluid filters through and through it as water through sand. The mucous linings show the same porosity as the skin, and when the latter becomes obstructed are often called upon to do the work which alone should be done by the skin.

Beneath the true skin, as we have seen, there are a great number of little sacs or follicles containing oil, deposited for the purpose of keeping the surface soft and preventing irritation from the acrid nature of the perspiration. This oil together with the dead cuticle and solid matter expelled unite, forming a gluey substance which as effectually closes up the pores as though the body had received a coat of glue. This refuse matter, if allowed to remain, prevents the exit of water, effete matter, and the gases, in which case the work of the skin must be performed by the kidneys, mucous membranes, and the lungs.

Consequently, catarrhal difficulty, bronchial affection, diarrhœa, leucorrhea and kidney complaint will be the result of this abnormal condition, as the waste matter must be expelled through these avenues in order to prevent death, which would as surely take place as though the individual were submerged under fifty fathoms of water. This system of internal drainage will go on until the outer gates are opened and a normal condition of the perspiratory system established.

We discover then that the skin is not simply a covering made to make the body appear more beautiful, but a complex and important organ, quite as delicate and varied in its structure as the lungs or kidneys, the cuticle serving as a thin, flexible, transparent media through which pass and repass impressions to and from the sensitive layer underneath, just as the senseless glass in your windows furnishes a means of communication between yourselves and the outside world.

In our study of the physical economy we are constantly impressed with the wisdom displayed in its design, and the protections used to guard its vital parts. All sensation, motion, and life arises from the nervous system, and those portions having the greatest supply of

nerve matter manifest the greatest sensitiveness. The
nerves can not come in immediate contact with gross
matter, and are therefore carefully enveloped and
guarded. For example, the real organ of vision, the
optic nerve, which receives all impressions, is guarded
on all sides by membranes and aqueous humors, and in
order that impressions may reach this sensitive plate of
nerve filament, we find in the front part of the eyes a
thin, transparent, oval membrane known as the cornea.
No crystal ever formed could compare with it in
thinness, clearness, and delicacy, but it is nothing
but a membrane, soft and flexible, and kept clear,
thin, and in a healthy condition by the tear-water
secreted by the lachrymal glands situated at the upper
and outer corner of the eye. The water is constantly
spread over the cornea by the effort of winking, which
is partially an involuntary impulse. If by any means
this movement of the eyelid should be suspended for
even an hour the sight would be much impaired if not
destroyed, as the cornea would dry and thicken, preventing the passage of impressions to the optic nerve.
This condition is often witnessed in dying persons when
the nerve controlling the eye-lid becomes paralyzed as
death approaches, and the lid is held stationary and
there is formed what is known as the "film of death,"
which is only a drying of the cornea, not affecting the

nerve of sight. There has been a blind drawn across the windows of the soul.

The cornea is just as insensible as the cuticle, and serves the optic nerve as the cuticle serves the ten thousand millions of nerve filaments that spread themselves over the surface of the body. Stretched across the vestibule opening into the ear is a membrane known as the tympanum or drum of the ear, performing the same service to the auditory nerve which the cornea does to the optic, or the cuticle to the sensory nerves. We find each of these insensible mediums supplied with a lubricator. The eye with water, the purest that can be eliminated from the organism, secreted by the lachrymal glands and constantly poured out to keep the cornea moist and clear. The tympanum has its oil in form of ear-wax, the oily portion being absorbed by the delicate membrane, which is thus kept clear and sensitive, so that the slightest vibration of air striking it at once produces an impression which is conveyed to the inner ear. But if through disease the lachrymal fluid or the ear-wax ceases to be secreted, the sight and hearing are greatly injured.

The lubricator intended for the epidermis or cuticle, evidently is the watery and oily secretions which in healthy persons are continually exuding from the pores. Let these avenues become choked up by dead cuticle

and solid cheesy matter of the perspiration, and the
membrane dries, hardens, and loses its sensitiveness,
and the communication between the electric life of the
air and the nerve filaments is lost, and more or less
irritation of the nervous system ensues.

In diagnosing brain and nervous disorders one fact
constantly presents itself to the studious practitioner,
namely, a total loss of action of the skin, the surface
being dry and feverish and the palms of the hands and
soles of the feet parched and lifeless. There is also a
peculiar sickening and offensive odor from the skin and
hair noticeable only in these instances. Many physicians are thus assisted in correctly diagnosing cases
from this aura alone.

USE AND ABUSE OF BATHS.

The first step in the treatment of any disease should
be to establish a healthy action of the skin, which may
be brought about by a systematic use of hot baths, with
a generous supply of soap and ammonia, together with
vigorous kneading, rubbing, manipulation and friction.
The body should be immersed as nearly as possible,
remaining from ten to fifteen minutes in the bath, so as
to soften the gluey substance on the skin, which an
ordinary towel or sponge bath has little or no effect
upon.

After thoroughly washing with soap, dip the hands into tepid salt water and rinse off carefully, then commence the rubbing and continue until the dead skin rolls off from every portion of the body. Rinse again with the salt water, dry thoroughly with a rough towel, then knead, rub, and manipulate the entire surface. There need be no fear of injury from this treatment. Rub and roll vigorously over the bowels and liver, under the arm-pits and over the thighs. Then annoint with vaseline, using but little, and rubbing it in until the skin glows and shines like satin. "He who keeps the skin soft and ruddy, shuts many gates against disease." is an aphorism well worth remembering.

Delicate, thin and nervous persons require oiling or annointing after each full bath, as the food consumed in most instances is of a nitrogenous character, instead of fat-producing, and little or no oil is secreted by the oil folicles, consequently the skin is not kept moist and flexible. There is a prevailing idea that fat meats, table oil, butter and cream are injurious and to be avoided by this class of individuals, when on the contrary—in the northern latitude especially—they are as absolutely indispensable as so much fuel or warm clothing. "Plenty of oil and honey on the inside and oil on the outside" was Pliny's recipe for longevity.

We are constantly hearing the plea that women have

7

no time for taking such baths. Now listen; it will
require one-half hour for such a bath as we have
described, and which need not be indulged in oftener
than twice a week if done thoroughly, and should be
taken upon retiring, as perfect rest after such a bath is
of the utmost importance. Then, too, it produces in
most cases a quieting effect, and occupies a time which
would not be used for other purposes. There is no
country in which the bath is so much of a necessity as
in our own, on account of the dryness of the atmosphere, which greedily absorbs the moisture from everything. This is especially true of the western portion
most remote from the sea-coast. This lack of humid-
ity in the air produces hardness and dryness of the skin,
which frequent bathing and annointing counterbalances
to a large extent.

Where individuals are too delicate to give themselves
this attention, there can always be procured healthy
persons who will perform this office for them at a trifling cost. Standing first in importance as a remedial
agent is the Turkish bath, which if within reach should
be resorted to in all cases of membraneous, skin and
glandular difficulties. It is a powerful stimulant to
the circulatory system. It relieves internal congestion by opening up the external avenues for the escape
of the poisons contained in the blood. When this

bath cannot be obtained, then home-made ones must be substituted.

In small towns, where the regular baths are not established, the ladies could club together, provide a room and furnish it with all the necessary appliances, employ a good, strong, *magnetic* woman to administer the baths and all the movements required, at a trifling outlay to each. It is no uncommon occurrence for an attendant in a regular establishment to give twenty baths a day. Each member of the club could have her regular days, so that there would be no collision and a large number accommodated, proving a luxury and a safeguard to those in health and a means of relief to those who are suffering.

The hostility manifested toward Turkish baths in this country is the result of ignorance on the part of the masses and want of candor, to a great extent, on the part of the medical fraternity. Every candid medical practitioner who has carefully studied the effects of the Turkish baths when properly administered, must acknowledge its supremacy as a curative agent over ordinary medication. It assists Nature without retarding her work. Heat is the radical of all cures. All fevers are simply fires kindled to burn out the intruders in the form of various poisons in the blood.

The heat cure meets a greater number of demands of

the system than any other one assistant ordinarily employed in medical practice, creating an activity in the entire cellular structure which any other single agent fails to do. The prevailing idea that Turkish baths are weakening is a fallacy growing out of a superficial knowledge of that system of bathing. Those people who have little or no action in the skin and lymphatics, when the liver and kidneys are overworked and diseased and the blood loaded with effete matter, will suffer more or less inconvenience from the first few baths, just as they would at first in the use of the various medicines. The bath arouses the dormant energies of the tissues to do battle against a common enemy. The first bath rarely leaves a comfortable feeling in persons with inactive skins.

The writer, while taking a series of Turkish baths at one time, had the opportunity of studying the cases of a great number of female patients taking similar treatment. In several instances as many as eight baths were required before the slightest moisture was perceptible upon the skin and two or three hot douches being required during each bath, in order to enable the patient to endure the intense heat and the consequent pain in the head. All of these patients were suffering variously, from menstrual suppression, leucorrhœa, catarrh, and nervous disorders. By perseverance in the treatment a

normal action of the skin was established and a permanent cure effected in most instances. Three-fourths of
the ladies testified to the fact that they had never
detected a particle of perspiration on any portion of the
body, neither under the arm-pits or on the face, even
when vigorously exercising or very much heated, until
after taking the baths for a time. One of the number,
a young girl who had been an invalid for two years from
some difficulty of the spine and general nervous system,
received two full baths daily for two weeks, and gained
in weight during the latter portion of the time a half
pound each day and was enabled to walk with ease and
comfort a considerable distance every morning, an exercise in which she had not indulged for over a year.

In most forms of heart disease these baths give almost
immediate relief, particularly when the difficulty arises
from defective circulation, producing an overcharging
of that organ and main blood vessels. The heat at once
creates an action in the capillaries on the surface of the
body, calling away the surplus blood from the interior.
When the capillary circulation has been defective for
years it will require a more extended course of baths to
permanently establish a normal circulation.

The body readily forms habits, the bathing calling
the blood frequently into the weakened and collapsed
capillaries will in time form a habit in that direction,

but one or two baths will not perform a cure any more than one or two doses of quinine would be sufficient to break up a severe attack of chills and fever.

All congestion and inflammation, such as throat and lung difficulties, the various eruptive fevers, small-pox, scarlet fever, measles, are greatly relieved by the baths, and the patient in nearly all cases at once removed from danger, providing the proper precaution is observed after coming out of the baths. There is no doubt but what much permanent injury and suffering has resulted from a neglect to guard the surface sufficiently when going from the hot rooms. A *sudden* transition from *hot* to *cold* should always be avoided, as a certain degree of cold closes the pores and contracts the capillaries, thus driving the heated blood from the surface to the internal linings. Persons suffering from irregular action of the heart and imperfect circulation should never be subjected to cold douching or plunges; indeed there are few who are strong enough to bear the shock of the plunge. Especially is this true in the cases of delicate women. Those persons suffering from fatty degeneracy of the heart must observe great caution in taking the heat at first, remaining in only a short time and drinking freely of cold water during the interval.

In cases of plethora or general dropsical tendency, the Turkish bath furnishes a desideratum of paramount importance. Patience and perseverance will be required in this branch of Therapeutics as in all others; where the strength has been for years gradually deteriorating the restoration will be correspondingly gradual, and when Nature has become exhausted by long continued abuse this agent must fail like all others. There is a limit to physical endurance, and when that is reached no power can restore the lost balance. When this fact is well understood, man will cease to recklessly squander his stock of vitality, knowing that he can neither beg, buy, or borrow a new installment.

CHAPTER III.

CIRCULATION OF THE BLOOD.

In order that we may more fully appreciate the great importance of a rapid and unobstructed circulation of the blood through all parts of the structure, we must first thoroughly understand the office of this wonderful fluid, and we will now briefly examine its composition and the manner by which the various changes are wrought in its passage through the body.

When examined under a microscope the blood no longer appears like a simple red fluid, but is found to be composed of two distinct parts; the first, a clear colorless fluid, richly charged with the material derived from the food, such as albumen, fibrin, fatty globules, sugar, salts of various kinds, acids of various kinds, in fact all of the materials that go to make up bone, muscle, tissue, etc.

The second part consists of a multitude of minute red and white bodies which float in the watery fluid; the white ones are larger, spherical in form, and not so numerous as the red, which are flat, oval, disc-like bodies, possessing great regularity of form, and so small that, according to some physiologists, it would require fourteen thousand to stand an inch high if they were piled up after the manner of coins, and in a cubic inch of human blood there would be several millions of these tiny discs. In these infinitesimal bodies resides to a degree the life of the blood, which is the great builder and renovator, as well as the purifier of the body. It not only conveys new material to the entire structure, but it likewise removes the wastes, worn out tissues and poisonous gases.

Blood is the fighting element, the red discs appearing like a multitudinous army, ever ready to do battle against an invading enemy, when it is not weakened and demoralized by enemies in its own ranks, in the form of various poisons.

In consequence of its being the vital principle of life, it must be conveyed to all portions of the organism, and for that purpose we find two sets of tubes, known as veins and arteries. The latter start from the heart, and pass outward toward the surface, dividing and sub-dividing, until they ramify every portion of the structure; these vessels are firm and deep seated, and convey the arterial blood to its destination.

The veins commence where the arteries leave off, and are connected with them by the capillaries, a set of

hair-like vessels through which the blood passes from
the arteries into the veins, and during that passage loses
its scarlet appearance. The veins lie near the surface,
the smaller ones uniting together forming larger and
still larger branches as they pass inward, until they
enter the body, where they combine, forming one grand
trunk, terminating in the heart. Thus we see that the
blood, in its passage through the system, describes a
circle; from the lungs and heart outward through the
arteries and capillaries, then through the veins onward
to the heart and lungs again; the muscular contraction
of the heart impelling the nutrient blood rapidly outward through all portions of the body; while in the
veins the turbid volume slowly passes inward, constantly
receiving new supplies of effete matter and poisonous
gases, which are thrown in by the numerous scavengers
situated along their course known as absorbents.

Now, in all of the efforts of life, thinking, breathing,
moving, etc., there is involved a destruction of brain,
muscle, and tissue, which being composed of minute
particles are little by little worn away in the friction.

These broken-down animal structures form poisoned
gases, and inflammable matter, which if not expelled
from the system would destroy life. This refuse is
emptied into the veins, and by this wonderful net-work

of tubing is carried through the various organs especially constructed for the purpose of receiving and expelling these poisons.

In the physical economy each organ or set of organs have their own particular work to perform, and in a healthy condition of the system no one will be called upon to do the work of another. For example, the kidneys secrete the urine. No other organ can effectually perform this office. The liver secretes bile, which the kidneys could not do, but the perfect law of adaptation is at once recognized, and the kidneys do their work and the volume is lightened of both solids and fluids; then the liver responds, and the blood is freed from bile, but the stream is still dark, there still lurks within its depths a deadly enemy, but as it rolls on, impelled by the unerring law of attraction, it is drawn into the veins that fill the lungs.

These veins branch out into tiny capillaries, just as the arteries did at their terminus. Passing downward into the lungs, there is a set of hollow bodies known as bronchial tubes, or air passages, and grouped about the terminus of each there are thousands of delicate membraneous cells or bags, whose office is to hold the respired air until the blood shall arrive to receive new life and to be freed from its burden of impurity.

As the dark lifeless stream comes in contact with the

air cells filled with pure air, that lurking enemy in form of carbonic acid gas leaps the barriers, rushes through the delicate membrane, and makes its escape through the bronchial tubes to the outside world, and in exchange the blood receives oxygen and ozone. An instantaneous change takes place, the blue color disappears, and there is a blossoming out of the red principle which characterizes arterial blood, and now, rich and warm, it passes into the heart, then goes surging outward through the countless arteries, carrying new life and strength to the most remote portions of the organism. Yet how comparatively little has been lost from its bulk,—this rich, red, buoyant stream, bearing life and strength in its depths, is the same that a moment ago was so heavily loaded with effete matter.

How noiselessly and perfectly this change has been wrought. Not only have the impurities been expelled, but as the volume nears the heart an ever watchful care lets in a rich stream of nutriment which has been elaborated from the food supplies, so that brain, bone, muscle, and tissue are fed and clothed withal.

By a wise provision of nature in this department, we find that like attracts like, and the worn-out particles are replaced by new, and thus a harmony is maintained.

The circulation of the "vital fluid" in man is governed by the same law as the circulation of water through the earth; the blood veins, absorbents, and arteries bear the same relation to the physical system that the drain-tiles, gravel-beds, and fresh-water pipes bear to a great city.

The filthy fluid in the sinks, cesspools, and sewers, rife with its deadly principles, would, if obstructed, breed all manner of pestilential diseases; yet if unobstructed, obeying the wise law of attraction which impels it toward the great rhythmical heart of Nature, the Ocean, it filters through soil, over sand and gravel beds, seeks the rapid flowing stream, dashes over the pebbly bottom, becoming cleansed long before it reaches its source; or, passing through the alembic of nature, ascending through trunk, branch, leaf, and blossom, falls back to us in rain-drops and dew, pure and limpid as when it came first from the great fountain.

Motion is the great law of life, equally applicable to animate as inanimate nature. Stagnation is death! mentally, morally, physically. The established laws of nature must be obeyed if we would escape suffering and obtain health and happiness.

We see by the hasty examination which we have given the subject, how important it is that the circulation of the blood should be rapid and unobstructed in its course. We discover that the veins lie near the surface, and are exceedingly delicate in their structure, are easily closed by pressure, so that clothing, either tight or heavy would tend to greatly retard the progress of the blood **toward** its sources of purification; and as there is a continuous column **passing from** the heart through the arteries and capillaries into the veins, we must at once **perceive that,** even with a slight obstruction, the adjacent parts must become overflooded and congested.

Fully three-fourths of the difficulties generally attributed to heart disease are the result of impeded circulation,—the blood not being permitted to pass and repass freely over its circuit, and in consequence there is an over-charging of all the large veins and arteries leading to and from the heart, the walls of which become unduly distended, the action accelerated and irregular, producing more or less suffering in the form of palpitation and fluttering of the heart, rush of blood to the head, causing dizziness and sense of suffocation, not the result of disease, but overwork on the part of that little-understood and much-abused organ.

There is displayed in the construction of the heart the most wonderful wisdom and economy. The muscular fibres are firm in their structure, and of that peculiar spiral form which enables them to contract and expand with great force and rapidity. It is an elastic bag which expands to receive the blood; then, in order to force that fluid **from its** cavities, contracts; this rhythmic action, known as systole and diastole, is unremitting in health. Nearly **all** bodily functions, with the exception of circulation, may be **for a** time suspended **and** life not destroyed, but the labors of the heart are ceaseless; **it** commences its work **in** the early stages of fœtal life and **continues** until death, even should the individual **live beyond the three-score and** ten; it is called **upon** during great **mental** excitement or physical strain **to perform a vast** amount of labor in an incredibly short space of time, hence the rapid beating **and** fluttering **of** the heart under those circumstances.

This organ complains less than any other in **the** economy, yet does **far more work** than all combined. It would seem incredible **at first** thought that so small a body could exert such enormous propelling power, yet science demonstrates that there is received and expelled from its cavities about eighteen pounds of blood each minute; **in** twenty-four hours, twelve **tons; in** one year, nearly four thousand tons. These figures indicate, in a measure, the immense labor performed by the circulatory system. The stomach growls incessantly if overworked, and kicks up a row **at** every unusual mental **tal** excitement; and the brain balks and gets "soft," so it is said, **if we** work it **a** little too **hard**; the extensor and flexor muscles cry out piteously if called upon to do an unreasonable amount of labor, and we are compelled to give them rest; but the heart keeps steadily on with its monotonous strokes, rarely reminding you of its existence. This insensibility on the part of this organ grows out of the fact that it is not largely supplied with sensory nerves. All organs in which are centered a great number of these are more or less influenced by mental conditions; mental disturbances, for the time being, preventing the performance of their normal functions; hence the wisdom displayed in the formation of the heart. That organ must work though all others refuse for a time. It is the first to commence and the last to cease, and the one that is the least liable to disease of the entire group composing the body.

If the heart was as much influenced by mental moods as is the digestive apparatus or the general muscular system, man would not live out half his days. The whirlwinds of passion which at times sweep over his being would paralyze the heart, and instant death would be the result; though there is no doubt but that the heart is to some degree influenced by bad passions, but not to the extent that other organs are which are more intimately connected with the brain.

Dr. Draper, who, owing to his careful investigation

of the subject, is perhaps the best authority on this point, says that out of every ten thousand persons that die, not more than one has any organic difficulty of the heart.

In cases of blood-poisoning, general relaxation of the system, indigestion, nervous irritation, and various diseases of the genital organs, we find that the heart, through sympathy, shows at times symptoms of serious organic difficulty; but as soon as these diseases are remedied, the action of the heart becomes normal and the aggravated symptoms disappear. It has transpired in the majority of cases that the post-mortem of subjects supposed to have died of heart disease reveals aneurism of the main arteries, but no organic disturbance of the heart.

It is estimated that there is a given number of heart-beats in an ordinary life-time: if this be true, then whatever would tend to accelerate its beating would to a degree shorten life. We believe that the present average longevity among civilized men is forty years, or thereabouts. In that time it is estimated that there would be about fourteen hundred million pulsations.

According to this calculation if the action of the heart is increased one-fourth then life would be shortened the same ratio. Great mental excitement, severe and protracted physical exercise, the use of stimulants

9

of various kinds, exposure to intense heat for too long a period, all tend to increase the pulsations, which varies in different individuals, the average in an adult being seventy or seventy-two to the minute, in very young children one hundred, and usually falling below the standard in extreme old age, the averages being much higher in the civilized than in the savage races. It is the generally received opinion that a full, firm, rapid pulse is a denotement of mental and physical activity, which is no doubt true, as a rule; still, we have cases in marked opposition to the above theory; for instance, those of Wellington and Napoleon, the pulse being not far from forty to the minute in the case of each.

We discover that the various positions of the body as well as food, exercise, and sleep, have a modifying influence upon the action of the heart. For example, if the pulsations were seventy to the minute while sitting, then standing would increase them to seventy-five or more, and by gentle exercise or a full meal, to eighty-five or ninety; but upon lying down for a time they would fall to seventy, and even to sixty, and during sleep they are a few degrees lower. This depression is merely a harmonious reaction, a rest, from which the individual rises refreshed and renewed in every portion of his being; but during great mental depression, exposure to cold, fasting and a suspension of physical exercise we find that the pulse sinks below the normal standard of health.

Two highly important facts here present themselves to the consideration of all who are in the pursuit of health and longevity. First, prolonged physical and mental excitation, over-eating and stimulating; in fact any course of living that will increase the pulsation beyond the normal standard in the individual, must tend to wear out, more rapidly than can be replaced, all the structures of the body, in which case there would be a febrile condition, accompanied by a loss of strength and great irritation of the brain and nervous system, a wasting of the flesh, an inability to sleep or to control the action of the mind, and the consequent shortening of life.

On the other hand we notice that a certain amount of mental stimulus, physical exercise, nutritious food and warmth is essential in order to give the necessary impulse to the blood and thus prevent stagnation and the ills growing out of that condition, such as heart difficulty, aneurism, congestion, local inflammation, etc., and if the theories advanced by Teigel, Billroth, Davaine and others, be correct, that the bacteria present in the tissues and fluids of the body are prevented from multiplying, collecting and preying upon the vital structures, as much through the rapid motion of the blood as through the integrity of the tissues, then we must readily understand how important to health it is that this fluid shall not be in any way retarded or obstructed in its round of circulation. And in order to establish an equilibrium and obtain the best results, both extremes must be avoided.

In one sense the body is like a machine, purely mechanical in its action, and subject to waste from friction just as any other machinery would be, but with this difference, that it constantly repairs its own wastes, unless the demands upon the system be greater than the supplies afforded. A machine that is kept in constant and rapid motion must wear out sooner than one used more slowly and carefully, while the one employed but seldom will be likely to rust out, and become useless even sooner than the one used in excess.

What Nature demands is action and relaxation. Periods of action must be followed by corresponding periods of rest. One is just as important as the other, and in order to preserve a harmony one must not exceed the other.

It is all important that we learn the lessons of letting down the tension on all of our faculties; of being able to rest at will; of putting off our pack as we lie down at night; of taking to our souls the assurance that this is not all there is of life, that we have still before us an eternity in which to labor and learn and enjoy; in fact, learn to wait with patience the fulfillment of the great law of our being, ceasing to fret because the acorn we have planted to-day cannot give us an oak to shelter us to-morrow. How infinitesimally small our cares and crosses appear when put in contrast to the grand ultimatum of life. They appear mere specks, and utterly insignificant. We must sleep more, and carp and fret and bicker less, if we would preserve our health and freshness, and keep away the wrinkles, which are not a necessary accompaniment of age, for Time can never bring to the face the disfigurement which would come through a disquiet soul; therefore, sleep away the wrinkles. Let us fold our hands, close our eyes, and let sleep baptize us in the name of Hope and Youth. Let us forget to be sordid and censorious, let us forget pride, ambition and selfishness, dreaming for a time only of the waking on that fairer shore. Let us grow young again.

The visible material body of the All-wise has periods of sound unbroken sleep; winter, ice-bound and silent, follows wakeful and jubilant summer—night follows day, cold follows heat and is its coadjutor.

A great thinker has said, "Motion and rest are the two wheels upon which the Universe moves." No rest is so absolute as that of sleep. It is for the time being

ruler of the realm.   It supersedes the rule of will.   It
is a wise and tender nurse.   It stands, with finger on
lip, at the entrance gate of consciousness, to hold the
passions in abeyance while the soul gains strength for
the coming battle; it sifts the experience of the waking hours; throwing away the chaff, leaving only the
grain in the soul's garner.   It presides over nutrition
and the general repairing of the system.   It is, indeed,
" chief nourisher in life's feast."

CHAPTER IV.

RESPIRATION.

WE must, by the previous investigation, thoroughly appreciate the importance of full, deep breathing in order that the air cells in the lungs may be filled to their utmost capacity with the life-giving principle; also that the respired air should be free from effete and poisonous matter, which is frequently of so attenuated a nature that the tissues of the air cells form no barrier to its passage into the blood, producing blood poisoning to a greater or less extent, as in the cases of epidemic and endemic diseases.

We discover this fact, that two bodies of equal density cannot occupy the same place, at the same time, therefore, if the atmosphere be loaded with moisture, dead matter, disease germs, etc., there can be but little space for oxygen and ozone. The minute molecular bodies composing the air occupy space, just as water and the denser mediums do; therefore, though the air cells may be filled with air, still if it lacks the life-giving elements, the impure blood cannot become properly arterialized and transmitted to the pulmonary arteries, but, instead, will remain in the veins, inducing various lung disorders, such as congestion, pneumonia, hepatization, etc., and in extreme **cases** terminating in consumption.

A careful examination **of the** confined **air** in most living and sleeping rooms leads to the detection of dust, refuse cuticle, disease spores, fungi, **wool** fibres from carpets, furniture and clothing, carbonic acid gas, carbonic oxide, especially where coal is burned, vapor exhaled **from the lungs and** skin; all of these deadly agents are being continually taken into the lungs **at** each inspiration, **but** not in sufficient quantities to destroy life **at** once; but enough to taint the blood and keep the inmates of **such** abodes **in a weak, half-dead** and alive condition, and susceptible to nearly every disease **the flesh is** heir to, especially to those attacking the mucous membranes.

We perceive **as we advance** with the study of these interesting subjects this all-important fact, which demonstrates the marvelous **wisdom** displayed in the construction of the physical economy, that **there is a** protecting power in the organism ever **on** the alert to ward off **all** destructive outside influences. This governing principle, doubtless having its seat in the cell structure, **causes the blood** to throw out various protecting insensible envelopes which stand between the vital organism and whatever would **tend** to destroy its equilibrium; **and so long as** the "vital fluid," through obedience **to** hygienic laws, is kept thoroughly cleansed and well supplied with **all the** materials necessary **for** repairing and building, then these outer walls, **which** guard the inner life, **are preserved** and kept intact, and health is maintained; **but** allow **the blood** to become **poisoned and** impoverished, and **the power to** form these barriers **is to a degree lost, and some order of** the genus bacteria, the **ever** present enemies to **human** life, take possession **of the citadel.**

In case of inflammation of the mucous lining **of the** nasal passages, throat, **bronchial tubes, æsophagus,** stomach, intestinal canals, etc., **we** find **that the protecting** tissues are not properly formed, **and the delicate** vascular structure is exposed to **the attack of these** creatures. **Such bodily** conditions as **we have mentioned** result **usually from one or more of the following** causes: Living **in dark, damp,** ill-ventilated houses, overshadowed **in summer by an overgrowth of** shade-trees and shrubbery, **and** darkened at **all** times by blinds and curtains, entirely shutting **out** the sunlight; or living over foul cellars that have defective drainage; **living** near marshes and sluggish streams that overflow **their banks,** and in **the** neighborhood of stagnant water, mill-dams, **etc.; by being** exposed to damp, cold

night air, with the body insufficiently protected by
appropriate clothing, thereby chilling the vascular surface of the body and driving the blood to the inner
membranes; or a neglect of the necessary bathing and
care of the skin; by taking too large quantities of food,
and of a quality that bears little or no relation to the
actual needs of the system; food unwholesomely prepared and taken at irregular intervals and at times when
the bodily functions are prostrated by physical or
mental fatigue; imbibing large quantities of cold
water during meals, thus chilling the delicate mucous
coat of the stomach and preventing the formation of
gastric juice, all tending to produce indigestion and
consequent inflammation of the entire alimentary canal.

In such a condition as this, the system is open to all
diseases of a membraneous character, as there would
be conveyed to the denuded membranes by the respired
air, water, and food, disease spores of various kinds;
and as a sequence hay-fever, acute catarrh, bronchitis,
asthma, putrid sore-throat, diarrhœa, dysentery, chol-
era-morbus, cholera, etc., will manifest themselves.
The leading scientists and microscopists have long been
satisfied that all endemic, as well as epidemic diseases,
such as diphtheria and the various forms of malignant
fevers, including yellow fever, black vomit, plague,
and, according to Dr. Hallier, scarlet fever, measles,

and small-pox, are the result of different **forms of bac-**
teria existing in the **air** and water.

If our dull eyes could see, unaided, the microscopic
side of creation, what wonders would be revealed to our
astonished gaze! We would see in our darkened rooms,
where light, sunshine, and pure air are not permitted
to enter freely, millions of infinitesimal bodies filling
the damp, impure air, moisture especially favoring the
propagation **of** fungi and bacteria; **hence** the importance **of having** perfect ventilation, light, and sunshine
in all habitations **designed** for man **and** beast.

Sunlight, owing to its power **to** dispel moisture, **is**
destructive to all kinds of fungi. These minute bodies
are no doubt being continually absorbed into the **blood**
through various avenues; but so long as all of the bodily functions are performed regularly and normally, and
the enveloping tissues **kept** unbroken **and** energized,
there can little harm **come to the structure** through
their presence. Tiegal says: "Rapid multiplication in
living bodies is prevented in part **by the rapid** motion
of the blood, **and in** part **by the vital** energy of the
tissues which is **so** vigorous that these spores cannot
check it, and thereby obtain nourishment needed **for**
their growth; but when **life has** ceased, or abnormal
conditions of the tissues has been brought about by any
cause, then rapid growth begins, and we have in one

case, putrefaction; in the other, various pathological changes of more or less importance.''

After careful investigation, Pasteur, the noted microscopist gives it as his opinion that the ordinary acute inflammatory complication of wounds, accompanied by symptoms of general poisoning, are due to the accidental entry of bacteria before the wounds are properly dressed and closed; and according to Davaine, malignant pustules are caused by a certain variety of bacteria.

There can be no doubt but that there are species which are very malignant in their natures. This is true in the case of larger and more advanced forms of animate life. Take, for example, certain varieties of reptiles: There is little difference between the general structures of the harmless milk and garter snakes and those of the malignant and venomous rattlesnake, adder, and cobra; doubtless certain of the bacteria possess the power to absorb from the surrounding media poisons more or less deadly, according to the nature of such surroundings. Whether these creatures simply convey the poisons inducing these diseases—as in the case of reptiles—or are themselves the causes of the various diseases, has not yet been fully demonstrated. Be that as it may, it is quite enough for all practical purposes for us to understand this fact: that any agent that would destroy the life or prevent the multiplication of the bacteria would tend to diminish the inflammation and thus assist in establishing a cure. There are various antiseptics that are wholly innoxious to the system, and yet have the power to destroy all protoplastic life. Two well-attested agents are borax and carbolic acid.

Recently a very interesting experiment has been made by M. Paul Bert upon the blood of a horse suffering from a disease known as glanders. On taking the blood from the animal there were found in it large numbers of bacteria in a state of great activity; dividing it he put in one portion a few grammes of pulverized borax, which at once destroyed the life of the bacteria, the blood remaining sweet and bright for days, the remainder becoming putrid in a few hours. This experiment brings fresh proof of the efficacy of borax as an antiseptic, and for the purpose of cleansing wounds and ulcers of various kinds, and for a gargle in all throat difficulties. In fact wherever the tissues have become abraded, broken, and inflamed, borax ranks high as a therapeutic agent. The antiseptic qualities of carbolic acid are so well known, and have been so thoroughly tested in the practice of our best physicians, that we need not more than mention it in this connection.

Another powerful agent is the cinchona in its various

forms; it seems to possess an extraordinary power to arrest the process of fermentation and putrefaction, and to act as a poison to all forms of fungi and bacteria when present in the blood; and all persons living under conditions that favor the growth of the disease germs will always regard this remedy as among the greatest vouchsafed to man. In this fight for life we must use the weapons that are most available and efficient, and least injurious to the physical organism, but always bearing in mind the old adage that "An ounce of prevention is worth a pound of cure;" and the only way by which to establish a permanent cure is to remove the causes that generate these destructive living bodies, remembering, also, that the physical system cannot bear the continued attacks without suffering a loss of power to do battle successfully with these multitudinous enemies. The ulcerated limb, although perfectly cured by the kindly agent employed, is never again so strong as before the first attack.

Diphtheria, although most effectually treated and cured, leaves the patient ever afterwards a subject to weakened and irritated membraneous linings, and to repeated attacks of the same malady. One suffering from diseased mucous membranes this season, will be more susceptible to a similar attack next season, as much perhaps from the force of bodily habit as from surrounding conditions, Nature not having had time to recuperate; then, **too,** perhaps all **of** the conditions have been against the recuperative effort.

What can be done? Remove all causes that tend **to** generate **and** perpetuate these enemies to life and **health.** See that the water which you consume is pure, untainted by animal or vegetable matter, that your houses are sunned **and** aired, **that your cellars are** drained, lighted **and** dried, **for nine out of every** ten **are** breeders of foul gases and disease germs of **a more** or less poisonous character; see that there are no cesspools nor foul and defective drains near your dwellings **to** poison the air, and, at whatever cost, do away with stagnant water. There can be but few greater curses to **a** community than **a** mill-dam in its midst, not alone causing **the various** derangements **of the** bodily **func**tions, but producing many important mental complications growing out of a poisoned condition of the blood. These cases will continue **to** multiply until the causes are removed or science has discovered some general disinfectant that will effectually destroy these pests of human existence.

Let Hygeia preside over your household, and sit at your tables **to** govern your eating and drinking. Let every house be supplied with a commodious bath-room, well equipped with suitable apparatus for administering hot and cold baths, and then use them often enough to promote a healthy action in the skin. Court the sunshine on all occasions, for it is one of the most important factors in the process of arterialization of the blood, exerting a peculiarly healthful influence over the red blood discs.

All living and sleeping rooms should be thoroughly aired and dried daily, using sun-heat when practicable, and in its absence artificial heat. Sleeping apartments require warming eight months of the year in this variable climate, more particularly in the fall of the year, and when the rooms are located in sunless quarters, as heat dispels poisonous gases, dries moisture, and prevents the generation of vegetable spores. It does, though not so effectually, what solar heat would do if allowed to reach the locality.

The first consideration in the selection of building sites and the arrangement of living and sleeping rooms, should be an unobstructed and sunny exposure; and when, through ignorance of sanitary laws, trees have already been planted too near the house, shutting out light, air, and sunshine, steps should be taken at once to trim or remove them entirely so that these healing ministers of Nature may have free access to every portion of the habitation. Have less solid wall and more apertures in form of doors and windows, and have the latter so arranged that they could be freed from the obstruction of blinds and curtains whenever it becomes necessary to air and sun the house. We are inclined to think that curtains and blinds in excess are inventions of sin, excluding all that is pure, and sweet and wholesome, and hiding within, many times, all that is dark and deadly. **Darkness,** if long continued, is destructive to all of the higher forms of animal and vegetable life; dark sleeping rooms, however **well** ventilated, **should** be avoided **as** we would avoid contagion.

Many **of** the severest cases of gangrene known **to** pathology have resulted from persons occupying **for a** length of time dark, damp sleeping rooms, those having **no** outside doors or windows. **The** occupants **of all** such apartments show blood poisoning in a more or less **aggravated form; the more common** manifestations being **an** eruption somewhat resembling salt-rheum, showing itself **particularly** about **the face** and head; **loss** of the red principle **in the** blood, tuberculous formation in the lungs, difficulty **in breathing, and a** general feeling of lassitude and nervous irritation, followed sooner or later by a total **loss of** recuperative power.

For example, **we will** give the case of **a** family whose **only daughter** had been an **invalid** for three years, and constantly **confined** to her room, which was on the north side **of the** house **and on** the ground floor. Dur-

11

ing the summer the light was excluded from her room by a dense growth of trees in the yard, and vines trained over the windows, making at mid-day an atmosphere of twilight in the sick room. The girl wasted day by day until she was a mere shadow, dying by inches, the parents praying hourly for submission to the will of God in their sore affliction. A new physician was called, though with little hope of benefit to the suffering girl.

Being a man of large experience and keen observation, he at once ordered his patient moved to a light, airy chamber, with sunny exposure, where on sunny days a sun-bath from half to three-quarters of an hour in length was given. A change for the better was perceptible at once, and in somewhat less than five months the girl was perfectly restored to health. Upon removing the furniture and carpets and mattresses from the north room where the sick girl had passed so many months, there was found a green mould on the baseboard, floor, between the mattresses, and behind the broken paper on the walls.

There should exist the best facilities for sunning and airing sleeping rooms, the bedsteads and all large pieces of furniture should be moved out, and the air disturbed and changed underneath them each day. The mattresses also require daily turning, drying and airing;

otherwise they will gather dampness, fungi will form, and foul gases be generated, and every time the bed is made or disturbed there will rise an impalpable dust, which the persons occupying or making the beds will constantly inhale, bringing about various disastrous results.

Through the aid of microscopy, there is constantly being brought to view so many startling facts connected with these subjects that the reading public in general must come to a realizing sense of the importance of the small things of life which ordinarily are passed by unnoticed, and in consequence of this growing knowledge the choice of building sites for dwellings will be more carefully made, the location of towns and cities will be a matter of deliberation based upon sanitary considerations, which if the founders of Rome had observed, that ill-fated city would have been the queen of the world to-day.

Houses will be built to live in, not merely to look at. They will be furnished with reference to the health and comfort of the inmates instead of style, showing little else than a lavish expenditure of time and money, and instead of wall paper and carpets, those catch-traps of disease germs and filth, we shall have painted and polished ceilings and walls, and hard-wood floors, all of which can be thoroughly cleansed and purified at will;

or, what will be better still, our walls, ceilings, and floors can be made of tile, which are elegant in the extreme, and owing to the increased facilities for manufacturing will **soon** come into general use.

In a chop-house in London there is a dining-room finished entirely in tile, the doors and frames of the windows being the only wood-work in sight. The panels between the doors and windows reach from the wainscoting **to the ceiling** in one solid piece, decorated with studies from the best masters, resembling exquisite painting on porcelain, having the appearance of the most elaborate and highly finished frescoing. This **entire** room—for the wainscoting **and** floor are also of tile—can be cleansed as easily as we would cleanse a dinner plate, and with no more injury **to** one than to the other.

How easily such a room could be cleansed in case of contagious diseases. **The** disease germs often remain in the carpets, curtains, **and wall** paper for months and even years, especially the germs of scarlet and typhus fever, small-pox, etc. Then what **a** saving it would be of bone, muscle, brain, patience, **and** Christian forbearance, **if we** could only escape that everlasting bore to men, **women, and** children, **that** semi-annual plague, house-cleaning, the tearing up, shaking, and putting down **of** carpets, what a saving of hard words on the part of men, and of broken backs, blistered hands, and soured tempers on the part of women, to say nothing of the blessed consolation of getting rid of swallowing such quantities of wool fibres, dust, and dead cuticle from the bodies of the unwashed, which are continually rising in the sweeping of our old, dusty, half-worn carpets, and constantly taken into our lungs and stomachs at every breath. Here the objection comes up, "But we have no hard-wood floors, and painted and polished walls." Get them then. They cost no more than your carpets and expensive wall papers, which must be quite frequently renewed if used constantly. A hard-wood floor, nicely polished, with the addition of a few handsome rugs, is a thousand times more elegant than the most expensive carpets, to say nothing of the durability and healthfulness of such a floor.

Any lady possessing ordinary genius can take a common white pine floor, have it smoothed, if it is not so already, fill up the cracks and nail-holes with putty, oil it, and give it two or three generous coats of shellac, and she will have a floor that she can be proud of. For the center of such a floor make a mat from matched widths of handsome carpet, bordered with rich, deep colors of a tone to correspond with the carpet or the decorations of the room, the size being varied to suit the taste. This rug should never be nailed down but frequently taken out of doors and shaken thoroughly, to do away with the necessity of so much dust being raised in sweeping, as well as preventing the wear on the carpet by the dust collecting and sinking into its fibres. Such a carpet would last five times as long as one nailed to the floor and removed once or twice in a year.

Then it is far more elegant than an ordinary carpeted floor, and it is within the reach of all who are able to have floors and carpets. All sleeping rooms, especially those used for sick rooms, should have these floors, which could be daily cleansed and disinfected.

Every home should possess a number of healthy and vigorous house-plants, which are of the greatest value as disinfectants and deodorizers, especially in sleeping apartments and sick rooms. The theory that plants exhale carbonic acid gas during the night has been exploded. Prof. Youmans recently tested the air, at night, in a room containing between six and seven hundred plants, and found it purer than that outside. It is a well-known fact that all plants inhale carbonic acid gas, which living organizations exhale, the plants in their turn exhaling oxygen and ozone, which sustains animal life.

"In addition to the pleasure which amateur floriculturists take in rearing their many colored plants in

garden plots during the summer, and sunny south windows in the winter, it will be satisfactory to those of
benevolent disposition to know that a learned scientist
pronounces them as benefactors of their neighborhoods,
and their cherished blossoms as a new class of physicians whose services are free to all and most effective
in curing many of the ills to which flesh is heir. It has
been known for many years that ozone is one of the
forms in which oxygen exists in the air, and that it
possesses extraordinary powers as an oxidant, disinfectant, and deodorizer. Now one of the most important
of late discoveries in chemistry is that made by Pro-
fessor Mantogazzi, of Paria, to the effect that ozone is
generated in immense quantities by all plants and flowers possessing green leaves and aromatic odors. Hya-
cinths, mignonette, heliotrope, lemon verbena, and the
whole list of our garden favorites, all throw off ozone
largely on exposure to the sun's rays; and so powerful
is this great atmospheric purifier that it is the belief of
chemists that whole districts can be redeemed from
malaria by simply covering them with aromatic vegeta-
tion. The bearing of this upon flower culture in our
large cities is also very important. Experiments have
proved that the air of large cities contains much less
ozone than that of the surrounding country, and the
thickly inhabited parts of the city less than the more

sparsely built, or than the **parks and** open squares.
Plants and flowers and green trees can alone restore the
balance ; **so** that every little **flower-pot is not** merely **a**
thing of beauty while it **lasts,** but **has** a direct **and** beneficial influence upon the **health of those** who care
for it.

CHAPTER V.

DIGESTION.

WE have seen in a previous chapter that the tissues composing the body were being continually worn out and as constantly replaced by new particles obtained from the food and drink, very little of which, however, in their natural state enters directly into the system, but by passing through the various changes they become sufficiently refined and attenuated to be received into the blood.

What entered the stomach as beef-steak passes into the circulation as albumen, fibrin, lime, phosphates, iron, sugar, oils and the various salts, substances readily appropriated by the system. The first step in the work of digestion is perfect mastication. By this effort the food is not only broken down, which greatly facilitates the process, but it is also mixed with the saliva, which alone has the power to digest starch and convert it into grape sugar, which can be used by the building forces, while starch in itself could never be appropriated by the system. Aside from the consideration of the chemical value of the saliva, there are several other important reasons why we should **eat** rather slowly and masticate our food perfectly.

First, more deliberation would prevent over-eating, which **is a** physiological transgression, for all supplies not required by the system for repairs and growth acts **as an** irritant and must produce **more or** less functional derangement. In the second place there would be less call for fluids during meals in order to rinse the food down. Third, the food **would be** more finely broken up, and therefore readily acted upon by the solvent juices **of** the stomach. Fourth, the masticatory act would tend to increase the flow of saliva, which **is, as** we have seen, an important adjunct to a healthy digestion, par-**ticularly** when starchy substances enter largely into our diet; **and** most of our breadstuffs **are** more than half starch. Rice is eighty per cent, and potatoes have little else besides starch to commend them as an article of food. And lastly, and perhaps more important than **all** other considerations, saliva has the power to bring out the flavors of food, thereby greatly aiding digestion; for the relish with **which** an article is **eaten** frequently determines its digestibility.

The second step in the process is the dissolving of the food after **it** enters the stomach. We find in the lining of that organ, innumerable tiny follicles or sacs which contain a clear, colorless, sour, and slightly saltish fluid, known as gastric juice. **As the** food touches the sensitive lining the muscular walls commence their peculiar peristaltic movement; the mucous coat changes from a pale pink **to** a bright scarlet, the temperature rises from sixty degrees to one hundred; **the** gastric juice begins to flow **out of** the follicles and mingle with the food, which through the solvent nature of this juice, **and** the ceaseless churning motion of the muscles of the stomach, becomes broken down **and prepared** for the third stage in this interesting process.

We observe that the stomach has two openings, one by which the food enters, known as the cardiac, from its proximity to the heart, being directly under that organ; the other, called the pylorus,—gate-keeper,—so named because of a valve which prevents the food not broken and chemically changed from passing out, and is situated at the extreme end of the stomach and opens into the duodenum or lower stomach.

Stomachal digestion gives us a brownish mass known as chyme, holding **in** solution the wastes as well as the nutrient principles of the food which is simply broken down. The gastric juice has had no power to digest fats or sugars, but as this mass passes through the pylorus **and** begins to accumulate in the lower stomach, the pressure provokes the flow of two other fluids **which** exert an important influence upon the partially digested food: namely, the bile, and the pancreatic juice,—the former a greenish bitter substance secreted by the liver, and the latter a colorless fluid with an acid reaction, secreted by the pancreas, a peculiar leaf-shaped gland situated just back of the stomach. There is an important ferment principle contained in the pancreatic juice known as pancreatine, which has the power to fit all oily substances for absorption, and now as these juices mingle with the food there is a separating of the nutritious from the innutritious elements, and a white milky fluid, called chyle, appears, which is the pure nutrient matter of the food. The principal work of digestion is now accomplished, and it remains for the lacteals to pick up and convey the rich stores to the blood.

The lower stomach opens into a long tube or canal familiarly known as the intestines, which varies in length, and is from twenty-five to thirty feet. About one-fifth of it is known as the colon, it being much larger, and serving as a drain by which to carry off the refuse matter of the food, as well as waste of the tissues. The walls of this canal, like the stomach, are composed of muscular bands that keep up a vermicular movement, slowly forcing the food forward. The lining in the small intestine covers about five times the space that the outside walls do, and upon close examination this mucous coat presents an appearance of plush, and each thread-like **body** when viewed under **a** micro-scope presents several little open mouths through which **the** nourishing matter of **the** food **is** absorbed, and **is** then passed through a beautiful network of small bodies, known as lacteal vessels **and** glands, and by this means it becomes still more perfectly filtered and fit **for** assimilation. The lacteal vessels unite, forming one large trunk, known as the thoracic **duct,** through which the chyle,—or **the** nourishing principle of the food,— is conveyed to the blood. In the continued worm-like motion **of** the small intestines, all portions **of** the food is presented to the open mouths of the absorbents; and here a wonderful integrity is displayed by the cell struc-ture **of** the villous coat: it absorbs only the nutrient particles, rejecting all waste **and** deleterious substances. Grouped among the absorbents in the lining, we **also** **find a** set of glands **which are** constantly giving out **a** fluid that not only lubricates the walls and facilitates the passage of the food, but also acts **as an** auxiliary in the digestive process.

By the time the mass of food has reached **the** colon all of the nourishment has been absorbed, unless the villous coat be diseased, and nothing remains but refuse **from** the food and worn out tissues. This should **be** expelled **from the** body daily, otherwise, as in case of constipation, the absorbents in the colon will pick the

waste matter up, little by little, and throw it into the
impure blood; that current quickly carrying it to every
possible outlet for expulsion, tainting all of the secretions more or less, giving rise to moth, pimples, freckles,
foul breath, indigestion, sick headache, and a long catalogue of ills growing out of the presence of poisonous
matter in the blood.

CONSTIPATION.

Constipation results from a variety of causes: among
the number may be mentioned lack of muscular exercise, of walking and full deep breathing, thereby bringing into play the heavy abdominal muscles, by which
means the intestines are constantly rolled and kneaded
and thus assisted in their work; the eating of concentrated food, an absence of a sufficient amount of
fruit and vegetables in the diet, irregular habits, and
a disregard for the promptings of nature, falling of
the bowels from wearing heavy clothing resting on
the abdomen, a dryness of the mucous coat of the
intestines, and a frequent use of cathartics. Many
others might be mentioned, but these we esteem the
principal causes for this distressing difficulty. Medicine
can have little effect so long as the abnormal condition
remains. A proper amount of exercise must be taken,
especially that of walking, the abdominal muscles daily

contracted through the effort of the will, rubbing, rolling and kneading the bowels, removing all bands and pressure from the delicate walls, making free use of fruit and vegetables in the diet, especially succulent articles as cabbage, potatoes, beets, turnips, onions, together with a full supply of good ripe sour fruit, and a generous quantity of rich cream and fresh sweet butter. Use coarse bread and oatmeal when they produce no unpleasant effect. Fresh stewed fruit, such as apples, peaches, prunes, should be in daily use.

A salad of some sort into the preparation of which good cider vinegar enters largely is almost indispensable to perfect digestion; as vinegar has a close kinship to the digestive juices, and when from some cause they are not abundant the vinegar proves a powerful auxiliary. Then each day, at a stated time, the individual must bring the mind and will to bear upon these functions. It is astonishing what a power they exert. Do not wait for a demand; make it by your will. Roll and knead the bowels with the cool hand; rub downward on the lower portion of the spine while sitting at stool; do not be discouraged if there is no action the first or even the second or third time; persevere, and success will crown your efforts. The bodily functions, even more than the mental, are governed by the law of habit. If you neglect to evacuate the bowels to-day the expulsive effort will be less vigorous to-morrow, and through this neglect very frequently the habit of constipation is induced. Another excellent remedy in case of a loss of tone in the structure is to take a daily cool hip bath, strongly impregnated with salt. It should be administered about two hours before the mid-day meal, remaining in the bath from eight to ten minutes, rubbing the abdominal walls vigorously during the time, neglecting at no time to center the will powerfully upon the desired object. A teaspoonful of white unground mustard seed taken before breakfast is an excellent assistant to the bowels. Another simple remedy is a tablespoonful of wheat bran wet up with milk or cream, and taken in the morning. When these agents fail injections must be resorted to in order to prevent the fæces from becoming impact in the intestines. The injections should be very warm, and copious, containing a little salt and just a sprinkle of capsicum. But above all medicinal agents in real value is the exercise of the *will* in this connection, together with the kneading and manipulation of the bowels.

EXPERIMENTAL KNOWLEDGE.

A knowledge of the digestive process has been obtained by actual experiment upon men and animals. Recently a case came before the medical men of Paris.

A boy swallowed by accident a dose of caustic potash. The terrible escharotic produced so powerful a constriction on the æsophagus, or gullet, that no food could pass into the stomach. Death from inanition must have been the result had not Dr. Verneuil resolved to perform the dreadful operation known as gastrotomy. Accordingly he cut into the lad's stomach, and inserted into the aperature an elastic tube through which food could be injected. •In this way, soup, fine-chopped meat, mashed vegetables, and drink were administered.

The young man recovered his health and spirits, and in one month gained ten pounds in weight, while being fed through the hole in his stomach. Of course this case must have a rare interest for all students of the complex phenomena of digestion, and it cannot but recall a still stranger one, which, nearly forty years ago, Dr. Beaumont, a Canadian, had the good fortune to observe. His patient enabled physiologists, in fact, to formulate nearly all our existing knowledge of the processes of stomachal digestion. He was one Alexis St. Martin by name, and, luckily for science, he had a hole made in his stomach through the sudden discharge of a gun. Dr. Beaumont cured him so far that he recovered his health; but, though the wound healed, the opening remained, and through it Dr. Beaumont was enabled to see the workings of a living human stomach in nearly

13

all conceivable circumstances. Another case of the
same sort—that of an Esthonian woman—came under
the observation of Drs. Schrœder and Grunewaldt some
four and twenty years ago.

Blondlot and others artificially produced the same
condition in dogs, and thereby gained a certain insight
into the behavior of food in the stomach.

It was found, for example, that in the fasting state
the walls of the stomach appeared pale and flabby, and
lay close together, whereupon some people erroneously
concluded that the sensation of hunger was due to the
rubbing of the coats of the stomach on each other. It
was noted that whenever food was introduced the sides
of the cavity reddened with the stimulated circulation,
and its muscular activity was aroused.

Of all the curious facts observed by students of the
phenomena of digestion, however, none was stranger
than the extent to which emotions were seen to affect
the operations of the stomach. Mental exertions pure
and simple did not seem to retard digestion so much as
had been conjectured. Yet when associated with deep
emotions or with that fretfulness of mind we call
"worry," it appeared to have a baneful influence. As
for the fiercer passions, an outbreak of them would
sometimes suffice to prevent the stomach, even when
most vigorous, from discharging its functions.

Another mysterious phenomenon noticed in connection with digestion **was** the sympathetic influence exercised by the stomach over the secretions of the mouth. This secretion, **as** we have seen, has an important chemical action on certain constituents of food, changing, by a sort of fermenting process, starchy matters, for instance, into sugar. **In Dr.** Verneuil's case, whenever food was injected into the stomach **of** his **patient "a** copious flow of saliva in his mouth **is** produced, in the ejection of which **a** motion curiously resembling chewing **is** remarked." In short, the different **parts of** the digestive apparatus are so linked together **by** nervous connections that they "sympathize" **with each other, and** if one be stimulated the rest **are also excited.** But, as might be expected, there **is** no connection apparent between the nervous system of the stomach and the nerves of taste, for it **is** related **of** Dr. Verneuil's patient that when food **is** put into his stomach, although his mouth fills with saliva, he feels no sense of flavor **in** the substance **with** which **he** is fed. He is only **aware** if they be hot or cold.

Cramped or stooped positions, long continued, **seri**ously damage digestion **by** compressing the nerves that **supply** the stomach. Those men and women who spend much of their time writing at desks, frequently suffer **from a** form of nervous indigestion. The case of R.

H. Newell (Orpheus **C.** Kerr) was of this description. For several weeks his consumption of solid food **did not** exceed a teaspoonful of rice daily, but he took several times during the day **a small** quantity of milk punch. **Mr. Newell** attended regularly to his business, being **on** the **staff of** the New **York** *World* during the period **of** fasting.   He suffered no pain, there being only a sense **of** repletion **as** though he had just eaten a **hearty** meal.

Dr. Carpenter gives account of several experiments made upon dogs.   When the pneumogastric nerve **was** severed the **animal at once** lost all desire for food, **and** ultimately died **of** starvation, while the most tempting **food was** being constantly urged upon **them.**

CHEMISTRY **OF FOOD.**

In the chemical analysis of **the human** body there are **found** a great **number of** different elements entering into its composition.   These principles are the products **of** the three kingdoms, mineral, vegetable, and animal. **Man,** being the epitome of creation, has **an** affinity **for** all that exists **in the lower** kingdoms; **the higher the** grade, the closer the affinity.   The albumen **and** fibrin obtained from meats are **more** perfectly adapted to his **needs than are those** found **in grains and** vegetables, having passed through a process **of** refining which better fits them **for** nourishing a higher type of organism.

The minerals and metals found in the physical structures are obtained from the different articles of food. If these minerals are taken in their crude state, they are rarely assimilated ; but after passing through those wonderful vegetable and animal laboratories, they are readily incorporated into the system, which in a healthy condition always contains a percentage of lime, phosphate of lime, soda, potash, magnesia, calcium, iron, sulphur, phosphorus, as well as the more sublimated extracts and essences of animal and vegetable food ; and in order that there shall be a requisite supply of all the principles in the blood, the diet must be varied,— perhaps not greatly so at any one meal, but a bill of fare ranging over a week should present a variety in animal and vegetable matter.

What is true of drug medicines of vegetable origin is equally true concerning fruits and vegetables of the food order. There is no doubt but what certain varieties have an affinity for certain portions of the organism, especially influencing those portions in the performance of their functions. For example, asparagus is a decided diuretic, not perceptibly affecting the system in any other way. The soporific influence of the onion is well known to every one.

A German scientist of note, whose name we cannot recall, gives it as his opinion that the hydrogenous prin-

ciple contained in cabbage supplies **a** delicate fluid for
the brain; he also **says that** the juice of the grape adds
to the serous fluid of the blood **a** principle not obtained
from **any** other article of **food, and** especially **recommends** that this fruit be given **freely to** convalescing
fever patients.   Many of our best physicians, and
among the number Dr. Hammond **of New York,**
recommends **in** cases of brain disturbances **and exces-**
sive nervous irritability accompanied by loss of sleep, **a**
liberal use of **celery in** the diet; and where this vegetable cannot be obtained, **raw onions** and celery salt **are**
to be used in its **place.**

BRAIN FOOD.

Every now and then **a new brain food is** discovered.
One day cheese is supposed to contain the needed material **for** reinforcing **and** strengthening that important
organ; then fish comes next, and then eggs; and there
is really no doubt but that the two last named articles
possess a large percentage of brain nutriment.   **Fish, if**
used **as** soon as caught, furnish a greater **amount** of
phosphorus than **any** other article **of diet, but** possess
little else that is of value as food.   The brain **does** not
perform its **work** well when meagerly supplied with that
material, and whatever article contains it most largely
will of course **rank** first **as** brain food, always providing

that it is both savory and wholesome. It must not be supposed for a moment that fish can make a good from an inferior brain. At best it can only help to sustain it. Food cannot affect the calibre of body or brain. Fish only acts as fuel, and is one of the many ingredients consumed by that organ, but is a very important one. Josh Billings is correct in his opinion, "That a man mite eat a whale and yit not hev enny branes."

It has been discovered by the French scientist, Bert, that the delicate oil found in the human brain resembles nothing else in nature so nearly as that obtained from the yolk of a fowl's egg; therefore it is assumed that eggs constitute a food for the brain, as a given amount of oil is absolutely essential to its health and well being.

After a careful study of dietetics, we are led to the conclusion that each article of diet has its own especial mission to the physical economy, no one taking the place of another, especially the vegetable products. Meat contains a greater number of constituent elements than any other article, and yet one could not thrive long upon an exclusive meat diet, and those subsisting entirely upon a vegetable diet are not properly nourished, and as a rule more subject to indigestion than those persons living upon a mixed diet. Flesh digests more readily than vegetables, with the exception of rice, which requires only about an hour in the process. The readiness with which the change is wrought is owing to its being so largely composed of starch. Those articles growing above ground, subject to the direct action of the sun, are more readily acted upon in the stomach than are those developed under ground.

Condiments occupy just as important a place in the well regulated diet as the food articles do. They assist digestion by imparting a relish without which the process could not well be perfect. It is claimed by physiologists that these relishes form no part of nutriment. We claim that all aromas, exquisite flavors and odors, as well as perfumes, do more than simply impart a zest to the gustatory act. They stimulate and at the same time feed that etherealized portion of our being which cannot absorb the grosser nutriment which our diet affords. It is claimed by many that alcohol is an article of food, yet they know that it makes neither bone, tendon, muscle, or tissue of any sort. The caffeine and theine contained in coffee and tea are not recognized as food, and yet we find that men allowed a reasonable amount of *good* tea and coffee have undergone severe and protracted labor without a loss of muscular strength, when their allowance of food has been less than that given in prison and army rations.

This matter has been thoroughly tested among

miners and soldiers **in** many portions of Europe, and our experiences in dietetics during the late war in this country also bear strong witness to the nourishing properties of coffee especially. **The** facts in favor of the **use of** tea and coffee while undergoing great physical and mental strain are further reinforced by the experiences of explorers who have been subjected to hardships, privations, **and the extremes in** temperature. Strain, in his expedition across the Isthmus, **bore** testimony to the benefit derived by his men, when deprived of food, in the use **of** tea **and** coffee. Dr. Kane also most thoroughly tested the efficacy **of** both. When his **men** were suffering from intense cold, and were nearly famished, these beverages would operate like a charm and lessen the desire for food.

We very readily perceive **and** understand **the** formation **of the various** portions of **the body as** well as the constituents composing **it, and** the food supplies required for building **and** repairing its structures. **We** understand **these things** because **we can** both **see** and feel in this department, **and** this constitutes believing; **but** we also know that there is an unseen power back of **all** this visible manifestation, and subject **to** analogous **laws. No one ever** yet saw the impulse of a thought along a nerve. **Who has ever** analyzed the nerve force? **Yet we know** that it exists. It eludes microscopic and analytic tests. Yet **if it is** force, it must be matter, though greatly sublimated, **and** must be constantly reinforced from **the more** spiritualized essences of the food **employed** in the diet.

DIETARY REGULATIONS.

It would be impossible to follow **arbitrary** rules in regulating **the diet of a** large number **of** persons, **for** articles of food which would **be** relished by one **indi-vidual would be most** distasteful if not injurious **to** another. Then, too, the tastes and habits change from year to year. For example, eggs may seriously disagree with one this season, but **just meet** the demands of the **system** and please the **palate next. A** great variety of elements enter into the composition **of** the body, and for perfect health **these must** be equally balanced. The loss of **one or the** predominance of the other will **dis-**turb the equilibrium. That marvelous instinct which resides **in** the cell structure **at once** detects that loss, **and** through the great pneumogastric nerve demands, through the appetite, an increased **amount of** whatever is lacking in the blood, and also **gives** the warning to abstain from eating such articles of food as are highly charged **with** the materials which the blood has in **excess.**

We, therefore, **have our days for eating** sweets, then

we crave sour and must have it; then comes a demand for salt; then certain kinds of meat only will do, then for days nothing but eggs will satisfy the appetite. Now we desire brown bread, then white bread only will be relished; last year strawberries were like poison to us, this season we cannot satisfy our craving for them, and nothing so perfectly agrees with us; we must have tea to-day but coffee to-morrow, and so on to the end of the catalogue. What may be wholly unsuited to us to-day will perhaps be just what the system requires to-morrow.

These variations in the appetite are usually called "fancies" or "notions," but this idea is not correct: it is the voice of Nature asking for those materials most needed for renovating and building up the system. The delicate nerves controlling the digestive organs are constantly sending to headquarters—the brain—their demands and protests in form of a desire or distaste for certain articles of food. The saying "Eat to live, not live to eat," was prompted by an understanding of the law governing supply and demand. The food consumed should be only proportionate with the waste of tissues. Eating was never intended by nature as simply a gastronomic pleasure to be indulged in regardless of the needs of the various structures of the organism. To consume more nutriment than can be appropriated

clogs the system and retards the work, as the overplus
is really waste, and will be expelled with the worn-
out particles; giving double work to the recuperative
powers. Occasionally a protest comes, in form of a
loss of appetite for all kinds of food, or perhaps only
for such articles as contain largely the elements not
required by the forces, the blood being already overcharged with those materials. Sick headache is usually
a vehement protest against overstocking the system
with nutritious principles.

The quality and quantity of food, as well as the frequency with which it is taken, should always be determined by the age, habit, and occupation of the individual. In early life the supply must be more generous
from the fact that material is not only required for
repairing waste, but also for the extension and upbuilding of the structure. In adult life only materials for
repairs are called for, growth being completed. In the
case of the aged, the waste of tissues is even less than
in middle life, and the food must be proportioned in
quantity, and suited in quality, to these various needs.
The brain worker requires a greater amount and of a
more varied and better quality than one merely using
the muscle, for it is estimated that the brain consumes
four or five times more blood than any other organ in
the body of the same size. It also requires a more del-

icate and refined matter to nourish **that** organ than **it** does **to feed** bone and muscle. **The** individual **who** uses neither brain nor **muscle** requires still less food, **as the** wastes of the system will be commensurate with the **labor** performed. Those persons making a free **use of** tea and coffee really need **less** nutriment than those doing without them, from the **fact** that the theine and caffeine principle contained **in** those drinks, **prevents** the waste **of** the tissues. Alcohol **has the same** effect, which accounts for the small amount of food consumed by inebriates.

Persons past middle life having little mental or phys-**ical** exercise, and yet eating heartily of rich food, drink-**ing** tea and coffee, must sooner **or later** suffer from **plethora** and its accompanying ills; such **as fatty** de-generacy of **the** heart, rush of blood **to the head,** vertigo, apoplexy, rheumatic gout, **hemorrhoidal piles,** varicose veins, **a loss** of muscular **power,** with a sense of fullness and discomfort generally. A repletion of the system is more **to be** feared **than a** degree **of** exhaustion in middle life; especially **in the** case of females ap-proaching the change **of** life, during which period an excess of blood is often **a prolific** cause of brain dis-turbance, from the pressure upon that organ.

It would **be** impossible **to give** a bill of fare which **would** please the taste **and suit** the needs of a large

class of persons, still we can give some rules of action
that will apply equally to all.

COLD DRINKS DURING MEALS.

We find that digestion is largely the result of fermentation, the ferment principle being contained in
both the food and the solvent juices. We also understand that fermentation requires heat to perfect the
process; it being suspended during the absence of that
important agent. Dr. Beaumont discovered in the case
of St. Martin that while the stomach was empty and
at rest the temperature was uniformly about sixty
degrees, but that immediately after food was taken it
arose to one hundred, at which point digestion was
most rapid and perfect. This rise in the tempera-
ature is occasioned by the influx of arterial blood in
the capillary vessels of the mucous lining of that organ,
and is called there for the purpose of increasing the
heat, elaborating the gastric juice, and stimulating
the muscular walls of the digestive apparatus. Anything, therefore, that would tend to lower the heat
would seriously interfere with the digestive process.
A glass of ice water taken at meal time would lower
the temperature of the stomach from twenty-five to
thirty degrees, the heat not being restored in less than
from three-quarters of an hour to an hour, and where

a weakness of the digestive organs **exists, the warmth** could only be restored through the .**use of** powerful stimulants.

The habit of indiscriminate drinking during meals **is** one that should be discouraged in children especially, although it is equally destructive to the health **of** adults. Copious draughts of cold fluid taken while eating, not only lowers the temperature of the stomach, but also dilutes and weakens the gastric juice, **in** which case the food would not be digested. Then too, during the time the liquids are being absorbed **by** the stomach digestion is suspended, and frequently the interval is of so long duration that the food becomes soured and loses its digestibility, giving rise to acidity **of** the stomach, flatulence, nausea, heaviness and pain in the gastric region, and many other premonitory symptoms of dyspepsia.

We must not be understood as discouraging the use of fluids. **They have** even a more important office to perform in the economy than **the** solids. In fact the greater portion of the body is water. It enters largely into all of the structures, bone, muscles, tendon, brain, **and** even the teeth, the densest part of the structure, **are ten** per cent water. Men can remain a much longer time without solid food than without water, but there **is a** time and **a place** for **all** things. Cold drinks, ice cream, ice cold jellies, fruit and melons, however ripe, are not admissible as a dessert at the close of a hearty meal that will require all of the heat and strength of the stomach to digest, but after the effort the organ is left heated and its muscles fatigued, just as the general system would be after severe and prolonged labor; then the cooling beverages, ripe fruit and ices would be most refreshing and beneficial, they requiring little or no effort on the part of the stomach to digest them. The watery portion of all food and drink is absorbed by the mucous lining, it being supplied with numerous open-mouthed, thread-like bodies, resembling that found in the villous coat of the small intestines. Thus we find the lining of the entire alimentary apparatus is governed by the law of endosmose and exosmose, and in this manner the serous fluid of the blood is constantly being reinforced from the water contained in food and drink; the watery principle of the blood acting as a conveyance for the nutrient matter and blood globules.

VITAL FORCE.

All organisms are endowed with a certain amount of latent heat or vital force, by which means the various processes of life are evolved. The percentage differs in different individuals, but the vitality required for a perfect process in any direction would be the same in all cases. It requires just as much vital force to digest an ounce of beef-steak in A's case as in B's. The expenditure in the elaboration of a thought would be the same in both cases; therefore, he who has small forces should not attempt to turn them in too many directions at one time, if he does he will fail in all—he may scarcely have force enough to make a successful effort in any one direction. We are constantly seeing these failures in imperfect digestion, assimilation, respiration, cerebration, etc.; in no direction do we find the work perfectly done; not because the organs engaged are at fault, but because there is little vitality, and that often recklessly squandered. The most important of all lessons to be learned is that of reserving the forces for the greater emergencies, and permitting nothing to interfere with or divert them from their course. A general who has a great army under his command can afford to send out large reinforcements in different directions without being materially weakened in his stronghold, but if his forces are small and he a wise and prudent commander, he will not dissipate or weaken them, knowing that his defeat would, in the end, be certain.

Franklin says "Attempt but one thing at a time, if you would do that thing well." Digestion and alimentation, taking them in all of their bearing upon the

15

system, are the most important of all the processes, and therefore should not be interrupted by physical or mental excitement. Every movement of the muscles or effort of the brain involves a loss of vitality in proportion to the effort. This vital force can ill be spared during the earlier stages of digestion, as there is rather a draft upon, than an addition to the vitality at that period. The immediate refreshment to the system after meals, comes from the fluids which have been absorbed by the coat of the stomach, and not from the solid food, which requires several hours to fit it for assimilation. We find after a hearty meal that the latent heat of the system becomes focalized in the hepatic region, often producing the effect of chilliness and generally a disinclination to exertion of any sort, and except where long continued exciting influences have created a feverish restlessness in mind and body, there will always come after a full meal an inclination for quiet, and in most instances, sleep, especially where the forces are weak. We notice an obedience to this law in the cases of young children, aged persons and animals, where the voice of instinct is not over-ruled by the will or the stronger passions. With delicate persons there will be after eating considerable excitement in the region of the stomach, a throbbing of the arteries, palpitation of the heart, especially after any exertion; this is simply the voice of nature admonishing the individual to rest for a time, and that admonition should be heeded and a short interval of sleep allowed. The *resting* may be of a longer period in the cases of weak or aged persons, the *sleep* should be only of a few moments duration—from five to eight minutes would be quite sufficient—a longer sleep frequently producing a feeling of exhaustion instead of recuperation.

During sleep a wonderful phenomenon is taking place in the system. The brain, that great consumer of vital force, is for a time quiescent. It lets go its vigilance—it rests—and the nerve power that is being continually consumed by it during the waking hours is now liberated and is being diffused through the body, those portions in action receiving it first. In the after-dinner sleep the stomach is the organ most in need of assistance, and the attraction will be in that direction for a time; then there is an onward movement and the outposts are reinforced, giving rise to that flush of heat over the surface of the body after one has slept for a few moments. It is this general dissipation of the forces that we should prevent. The sleep should be brief but the period of rest may be prolonged *ad libitum*. There may be some pleasant occupation that is not laborious, reading, or light needle-work, engaged in for the first hour, after which a reasonable amount of labor cannot adversely affect the process of nutrition; indeed, would rather aid it than otherwise, but it is during the sleep of the night that the great work of upbuilding goes on most perfectly—the food has been converted into chyle and poured into the blood, the little builders are active in giving up the worn-out particles and taking in the new; there is a conservation of all the forces. Those persons deprived of a proper amount of sound sleep soon show evidences of impaired digestion, for which medication of the best kind can do little, so long as a loss of sleep is sustained.

We cannot refrain from giving here Mrs. Elizabeth Cady Stanton's views upon rest and sleep, as expressed by that lady in a recent conversation. She said:—"I am a great advocate of sleep. I always believed in it, and I never allow anything to disturb me from sleeping. I have taught myself to go to sleep whenever I am tired. My old father, who was a well known lawyer, outlived four generations of young men who took no regular rest. Whenever he felt tired he went to sleep. At eighty-four he sat as a judge on the bench, with faculties clear as a bell, and with health unimpaired. Whenever he had an important brief to prepare he did not pace his room all or half of the night, but rested himself by refreshing sleep. 'Why,' he used to say, after he had gotten up after a brief nap, 'I can look through a millstone.' The mistake our business men make is that they do not take enough rest. Whenever a man is tired he should go to sleep. I would advocate a nap of from ten to fifteen minutes after every meal. If I had the power, I would put a lounge in every office, and compel every man to lie down at least half an hour each day. The rest on a lounge after meals aids digestion, and the horizontal position assists in the circulation of the blood, making it uniform throughout the body.

"If men followed Nature's laws more closely there would not be so many of them comparatively young filling premature graves. A short rest and nap is as exhilarating after a meal as a glass of champagne. The trouble is that men chew and smoke tobacco and cigars, and drink liquors, and get up a false stimulation, while all the time their constitution is being undermined."

INFLUENCE OF THE PASSIONS UPON DIGESTION.

None of the bodily functions are so easily influenced by the mental moods as those of digestion. The stomach is almost wholly under the control of the sentiments and passions. Anger, jealousy, hate, envy, fear, grief, each have the power to adversely influence the action of that organ, while on the other hand all pleasurable emotions assist it in its work.

Dr. Beaumont discovered in his experiments upon

Alexis St. Martin that a fit of anger for a time entirely suspended the digestive act, and that for many hours after there would be more or less functional derangement of the stomach, accompanied by headache and loss of appetite; while on the other hand, pleasant surroundings and cheerful conversation facilitated the operation always; and yet this soldier was not of an imaginative turn of mind evidently, else he could not have borne these long-continued experiments without failing utterly in health.

That the imagination has a very powerful influence upon the stomach, no one can deny that has been at all observant. Many sensitive persons will at sight of a pill become nauseated and experience all the effects of having taken a thorough purgative, and the mere thought of a lemon will cause a copious flow of the saliva and a peculiar action at the pit of the stomach. It is said that the men accompanying Strain on his Darien expedition, when destitute of food and nearly famished, would sit down together after a day's fatiguing march, and one after another of them describe sumptuous dinners, dwelling with glowing terms upon the various dishes, each trying to vie with his fellow in the luxuriousness of his imaginary repast, and for the time the appetite would seem satisfied and the hunger appeased.

At times the mere thought of anything unpleasant or unsavory will instantly destroy the appetite, and with it the power to digest food for a time. An unkind word spoken to a sensitive person while at table will produce the same unpleasant effect. This is especially true in the cases of children. All persons endowed with a sensitive nervous organization, possessing a keen perception of the fitness of things, will be affected more or less seriously by unpleasant surroundings, and for the well being of such, so far as the matter of health is concerned, order, neatness, and a degree of ceremonious courtesy must be observed. One with æsthetic and epicurean tastes requires for perfect digestion a delicate blending of flavors, without which the best articles of food would be most unpalatable and in consequence indigestible. To those possessing this fine appreciation the preparation of food is of the first importance. The entire alimentary process is greatly affected by the gustatory act. For such there must be an imprisoning of all the delicate juices and subtle spirits of the food which not only increase the nutritive value but delight the palate also. These apparently trivial things have an important bearing upon the health of persons thus organized. No one is to blame for this condition; they can no more help their peculiarities than they can change the color of their hair or eyes.

Another class of individuals will live and thrive upon a plain and indifferently prepared diet served in the most slovenly and unceremonious manner. To such persons flavors are of no account; they could never detect the difference between one dish and another; they digest all with equal readiness. The imagination has nothing to do in influencing the lives of such people; they are just as well satisfied with black, heavy, sour bread, strong butter, muddy and flavorless coffee, soiled table linen, defaced delf, as they would be with the most faultless surroundings. The extremes of both classes are to be pitied. But it must not be supposed for one moment that the outcome of these two conditions are the same, for man is essentially the reflex of his diet and surroundings. From the French epicure down to the clay-eaters we see demonstrated the fact that man is an outgrowth of his feeding ground, the dietetic influence reaching out through every portion of his being and shaping him physically, mentally, morally and spiritually. The difference between the organism of a truly æsthetic epicure and a common gross feeder is as great as that existing between the finest porcelain and the coarsest pottery; but more particularly does the diet manifest itself in the intellectual progress of man.

The Esquimaux never rises mentally or physically above the level of seal flesh and blubber. The Indian becomes civilized in proportion as he becomes a civilized feeder. The versatility of a people depends largely upon the heterogeneousness of its diet; and those nations who consume large quantities of meat show a marked superiority over those living exclusively upon fruits, vegetables, and grains. A great student of men and things has said, "Beefsteak swears outright in the physical organization, still without it there would never have been elaborated any great commercial schemes. Railroads, steamboats, telegraphs, etc., are not the enterprises of the vegetarian." The meat-eating nations are the ones that make the most rapid progress in all directions. The rice-eating Hottentot is to-day the counterpart physically and mentally of his remote ancestors.

All food has a soul or spirit, a principle not visible to material eyes. Who ever saw a perfume, an aroma, or a flavor? Yet we know that they exist, and for the well-being of a cultured organism are quite as important as the more solid portion of the food. Every perfume and aroma that gives delight to the senses has to an extent refreshed, stimulated, and nourished the most delicate and subtle portion of our physical being, the nerve filaments. They have penetrated beyond the domain of crude solid matter. They **are to** the nerve filaments what albumen, fibrine, casein, phosphate, and lime, etc., are to muscle, tendon, tissue, bone, and brain. In the preparation of food these subtle principles are being continually freed by the heat used **in the** process,—more noticeable perhaps in the preparation of meats, tea, coffee, **and** highly flavored dishes and confections, etc.

As a rule cooks and butchers are the best nourished **class of** persons, and yet are usually light eaters. **It is** in **the** preparing and **seasoning** of various dishes that the cook inhales and absorbs the delicate essences and aromas which are the real **life of** the food. The butcher receives vitality from handling **his meats, and** magnetic **life** from coming constantly in **contact** with living animals.

WHO SHALL DO OUR COOKING?

There **is often a** rapid recovery to health and strength in the cases of delicate ladies **who have taken the** supervision of their own cooking, particularly the seasoning **of** the food. The agreeable **flavor from** boiling and roasting meats, and dishes highly seasoned with aromatic spices, imparts **a** most delightful stimulus **to a** delicate organization, and if our **frail, sickly** girls, now under **the** care of physicians, were properly instructed **and** interested in that most fascinating of all **arts, cookery, and** were encouraged to give daily a certain portion of time **to the** preparation of **food,** they would soon manifest not only more robustness of physique, but also its sure accompaniment, a more harmonious mental development. There **comes a two-fold** blessing in the performance of this portion of household labor: **one,** in the absorption of **the etherealized essences of the food** while working over it; the other, in the gentle **exercise of** mind and **body** assisting **as they** do all of **the** bodily functions. No girl's education is complete where a knowledge **of this** branch of chemistry has been omitted.

Cookery in **fact** is not only a science, but a fine **art** also; and to understand **it in all its** branches requires **as** much intellectual ability **as does** any other department **of** science or art. According **to** Ruskin, "Cookery means **the** knowledge of the Queen **of Sheba, of Calipso,** of Helen, **and of** Rebecca. It means **a** knowledge **of** all **that** is savory in field **and** grove. It means much tasting and **no** wasting; **it means** watchfulness, patience, perseverance, and **tact; it means** English thoroughness and French art; **it** means Arabian hospitality and the economy of our grandmothers."

The great secret of perfect cookery is in preventing **as far** as possible the escape of the aromas and **flavors** from food **during** its preparation. Coffee that has regaled the nostrils of the household and neighborhood while being prepared, is of no further use, as the only portion that could be of any value to the partaker has escaped.   It is the volatile principle that must be retained if we would derive any benefit from the use of the beverage.   A high degree of heat long continued liberates the subtle essences contained in both tea and coffee, without which they are worse than useless, as in the boiling process there has not only been a disengagement and loss of the caffein and thein, but there has been extracted from the woody fibre of the berry, tannin and tannic acid; articles unsuited to the needs of the system and frequently producing indigestion, constipation, and headache,—difficulties which the well prepared articles in most instances would relieve.

NATURE'S RESOURCES.

In Nature the law of fitness rules all things, and when we obey this law, health and harmony prevails. The vegetable products are supplied to us at the proper time; each zone presenting to its inhabitants those articles which most perfectly meet their needs.   In our climate, during the months of June and July—our hottest seasons usually—we have the strawberry, currant, pie-plant, and other acid fruits which are cooling, cleansing, and laxative in their nature.   If these articles are eaten in excess, which they are quite liable to be, diarrhœa, dysentery and other stomach and bowel difficulties will make their appearance. At this juncture Nature comes in with her arms laden with remedies, in form of blackberries and raspberries, both black and red, and the earlier varieties of pears; all of these fruits being astringent and healing in their nature. Still later in the season, in what may be known as the typhoid season, we have in the vegetable line the tomato and onion, both of which, if eaten in generous quantities, will, as a rule, prove a safeguard against that distressing disease.

In case of dysentery, which is often a precursory symptom of typhoid, the juice of the tomato simply stewed with the addition of only a trifle of salt may be given to the patient in small quantities when all other articles in the food line would be impracticable, and there is often a strong desire for this article of food on the part of persons convalescing from these diseases.

During this season, as there is often a disposition to wakefulness from the feverishness of the system, a liberal use of onions will generally bring relief. A ripe onion eaten with salt and a bit of bread and butter just before retiring, will, in most cases induce sleep when all else fails. Aside from the soporific effect of this vegetable, it acts as a powerful stimulant to the coat of the stomach and liver. As an article of food it should appear upon the table in a cooked form three or four times a week. The fastidiousness manifested by many persons regarding the odor of onions on the breath is largely affectation, for so far as disagreeableness is concerned there can be no comparison between an onion and a tobacco breath; the latter being many degrees more intolerable to a fastidious taste. There can not be found a more critical and truly fastidious people than the French, yet the use of the onion and garlic is universal among them. There is scarcely a savory dish prepared that has not a hint, at least, of these vegetables. In this, as in many other things, it is a matter of education, and the appetite should be trained to like that which would be most essential to the system. If children were trained to the use of onions in generous supplies there would soon be no demand for worm medicines, and in the adult its use would, in most cases, correct the habit of constipation and remedy foul breath. They should be eaten either raw or carefully boiled in milk with little or no butter added, as that article is often the only thing in case of weak stomach that makes the vegetable indigestible.

Eggs as an article of diet occupy an important place in the bill of fare; but it is of the utmost importance that they be *fresh*, for the shell of an egg forms no barrier to the ingress of the vibrio, a species of bacteria, and they therefore become unfit for food if left long in damp places. They should be kept perfectly dry and in a cool, airy position; in fact, where a current of air can pass over them. Soft boiled eggs usually digest more readily than those which are hard boiled, or fried. Perhaps the most delicate manner of preparing this delicious edible is by poaching it, particularly if prepared for invalids, or people with impaired digestion.

In the autumn after cabbage has ripened it should have a place of honor among vegetables on every table; not cooked, but raw, dressed as a salad. It requires nearly five hours for the healthy stomach to digest boiled cabbage, while the raw vegetable takes but two hours at the outside. The facility with which the raw vegetable is digested arises from two causes: first, its porosity, which permits the gastric juice to enter its fibres and thus assist in breaking it down; second, from the ferment principle which it contains, and which the heat in the cooking process totally destroys.

Tomatoes also serve as a delicious salad, and should be in daily use during their season, not simply because they are delicious, but because they possess a medicinal virtue of great value which no other article contains.

There is constantly in reserve in the vast storehouse all that man needs to preserve his physical equilibrium.

It requires only a quickness of perception on his part to see and accept these gifts when held out to him by the gracious and loving hand of mother nature.

Milk is too rich to be used as a beverage with other nutritious food. It is in itself virtually solid food. After being received into the stomach it is converted into a sweet curd precisely like that of which cheese is made. This change is wrought by the pepsin contained in the gastric juice, and after the formation of the curd it is digested just as any other solid food would be. It is too hearty to be used as a drink for young children at night, usually producing a feverishness and restlessness occasioned by its stimulating nature. Then, too, the quality of the milk is of the first importance. The investigations at different periods into the milk traffic of London, Paris, New York, Chicago, and other large cities have brought to light some startling facts concerning the nature and composition of milk as affected by the food, drink, and surroundings of the animal.

In the recent microscopic investigations in Chicago there were discovered great numbers of disease spores in all of the milk obtained from the "still-fed" cows. These spores pass directly into the circulation of those persons using the impure fluid, developing serious difficulties. Especially is the effect disastrous in the cases of children,—infants more particularly. Cows should never be permitted to drink from stagnant pools of water, which are always impregnated more or less with organic matter giving life to millions of infinitesimal creatures of the bacteria species. These pass into the blood of the animal, then into the milk, as a matter of course, and then into the human organism.

M. Porville, of Paris, in a recent scientific work, speaking of organic matter in water says: "How does organic matter become dangerous? We must not believe that it constitutes, as is superficially said, a toxic element. The phenomenon is more complex. The organic matter in suspension or in solution creates in the water a peculiar medium, suitable for the development of exceedingly small beings of the genus *vibrio*. It is no longer mere water,—it is a world of microscopic animals and plants which are born, live, and increase with bewildering rapidity. The infusoria find in the water calcareous, magnesian, and ammoniacal salts, and their maintenance is thus secure. Drink a drop of this liquid and you swallow millions of minute beings. But there are vibrios and vibrios. There are those which are capable of setting up putrefaction in our tissues. These are our enemies, often our mortal enemies. Let water be placed in contact with organic remains capable of nourishing these malignant vibrios,

17

and it at once becomes more dangerous than any poison."

These considerations concerning pure water are more applicable to man than even to beast; but man has the power to improve his environments, which the animal has not. A cow will drink from any source, however impure, and therefore must be carefully guarded—her drinking fountain must be supplied with fresh running water as pure as that used by man. It is the small things that make up the sum of life, and nowhere do we notice the effect of small things so much as in the diet, and its complex influence upon the physical, mental and moral nature of the human being.

PREPARATION OF FOOD.

There is a lack of one desideratum in our regime, and that is, carefully prepared soups. The diet is incomplete without them. This dish when scientifically prepared is already half digested. The solid food entering into its composition has been, by a slow process of heat, broken down in a manner somewhat analogous to the process of chymification. It can be readily digested and assimilated; in fact, it is largely absorbed by the stomach as soon as taken, at once nourishing the system. The solid matter composing the soup would have required from four to five hours in the digesting, consuming heat and force which might be required in other directions; and from this consideration this dish is of the greatest value to students, teachers, and clergymen, in fact all persons having a given amount of mental labor to perform in a limited time, during which there should be no distraction of the brain forces. A dinner requiring four or five hour's labor on the part of the stomach must interfere somewhat with the mental processes, or *vice versa*. Whichever is the strongest will gain the ascendancy, but there must be a loss somewhere.

Soups should contain largely the rich red principle of the meat, the osmazome. But little or no oil should enter into the composition, as that is usually the portion that distresses persons of delicate digestion.

Roasted and broiled meats when they are delicately browned and nicely seasoned, are preferable to those that are fried, as all fats that have been subjected to a high degree of heat have been rendered indigestible, especially so to all invalids. The evidence gained by recent microscopic investigation into parasitic life in animals, would be in favor of well-cooked meats, as a high degree of heat destroys the life of these minute bodies. Particularly should this precaution be observed in the use of pork, as trichina is sometimes found in the flesh of that animal, which comes through the fault in the feeding and not through the animal itself, as the superior breeds which are well fed and kept cleanly are entirely free from these disgusting pests. The wholesomeness of all flesh food depends much upon the food and water with which the animal is supplied. Eggs and the flesh of fowls are influenced from the same causes, still these things do not affect the animal organism as much as many are led to suppose. All animal and vegetable matter after being digested by the animal passes through a sort of crucible, the absorbing system, by which means the original elements of the food have been stripped of their foul environments and enter the blood in a pure form, something after the manner by which glycerine is obtained. That clear, inodorous, sweet fluid is extracted from the fats at soap factories; these fats often being of the most fetid nature, yet the glycerine is as pure and sweet as honey, owing to the refining process through which it passes.

What we have most to fear from the use of flesh food is the infinitesimal parasites and their germs which the animal takes in with his food and drink and which the digestive processes have no power to destroy, they becoming in time incorporated into the muscular tissues, and if these are not destroyed by heat in cooking they will reappear in man, as they have at times, causing great suffering. Therefore, as a precautionary step, let all the meats be well cooked ; not even making use of rare fresh beef, the kind of meat least liable to be affected by parasites. Mutton, veal, pork, and fowls of every description should be most thoroughly cooked.

In roasting and broiling of all cut meats, the first heat must be intense, in order to sear over the outside, close the pores and prevent the loss of the juices, as well as the hardening of the fibres, which would render it to a degree indigestible.

In order that the muscular walls of the digestive apparatus may be kept in a healthy condition they must have a given amount of labor to perform at regular intervals. The diet therefore must contain a large percentage of waste matter, which will tend to excite the peristaltic movement, attracting arterial blood to the lining, increasing the heat and consequently the flow of the solvent juices. If the food is of a concentrated nature and only a small amount taken at a meal, the walls of the stomach are not expanded, only a small quantity of blood is called to the mucous membrane, an inadequate supply of gastric juice is secreted, the appetite fails, and indigestion follows. Therefore a reasonable amount of coarse food must be used, brown bread, oat and corn meal mushes, together with an abundance of fruit and vegetables of all kinds, varied from day to day in such a manner as to present a freshness and change which would be both appetizing and healthful, and although one might have a predilection for, or an aversion to, particular articles of diet, which should be regarded, still the appetite can be to a great extent cultivated in favor of different dishes and the diet varied thereby.

DIETETICS FOR CHILDREN.

The training of the appetite and taste should receive careful attention in the cases of young children, who after the teeth are perfectly formed should be gradually taught to eat, in moderate quantities, all wholesome articles of food. It is unwise to indulge a child in the fancy that it cannot eat this, that, or the other thing. The peculiarities of appetite are usually the result of a mere whim, the individual perhaps never having tasted the article toward which he has such an antipathy, and frequently after being induced to do so ever afterwards entertains a great fondness for that particular thing. Children are often encouraged in their peculiar fancies with regard to eating. One will not eat any kind of fruit; another cannot be induced to touch vegetables of any description; one perhaps living entirely upon bread and butter; another will be satisfied only with meat, while another will not touch it at all.

Scarce any of these children present a normal development, generally showing defective bone formation, particularly in the case of the teeth.

There came recently to our notice the case of a girl eleven years old, whose diet had always been exclusively bread and milk. Her mental and physical calibre is that of a child of seven years.

Much has been written, and still more said, upon the subject of dietary regulation for children; vegetarians advising the exclusive use of farinaceous food, and for years we were unwilling to accept any other doctrine; but through observation and careful study we have been compelled, however unwillingly, to admit that those children who were furnished with a generous and heterogeneous diet were—where the food was properly prepared, of a good quality, and given at regular intervals—the best developed.

The child is subject to the law of attrition and nutrition which governs the adult organism, and its body is composed of kindred elements. After the teeth are well formed there can be no doubt but that a mixed diet, properly prepared and intelligently administered, is best adapted to the well being of the child.

Stomach and bowel difficulty and so-called worm indications in children are more frequently the result of
overloading the stomach by eating between meals than
from the mixed quality of the food.

It is not advisable to indulge young children in
the use of tea and coffee, as they do not require the
stimulus contained in them which is so grateful and
beneficial to older people. In childhood there is a
positive condition of the system accompanied by an
accelerated pulse, ranging from a hundred to eighty or
ninety to the minute. Nothing should enter the diet
which would tend to increase the pulsations. Only the
most substantial and nutritious food should be given in
order to lay a firm physical foundation which may, in
the future, bear the wear and tear of active life.
Whatever would excite the nervous system would tend
also to prematurely develop the brain, which to a
mathematical certainty would detract from bone and
muscle formation. Discourage the use of cold water
at meals. If a child insists upon some sort of drink,
resort to cream and sugar, and enough hot water to
make it palatable. If the food is warm this may be
dispensed with, and in case the food is cold the warm
drink would assist in digesting the meal.

A lady in our hearing recently remarked that she had
tried in every way to persuade her children not to drink

so much cold water at the table, but **that she could do** nothing with them, **as they** would have it, drinking two or three glasses at a meal. Our suggestion would be, keep the **water** from the table. Many times the sight **of** it prompts the desire for it.

It **is not** only unwholesome, **but to an** extent vulgar, **to** drink inordinately during meals. Breeding manifests itself **sooner and more** unmistakably here than anywhere else. We often meet people whose manners while eating **are so** disgusting that those with **delicate** appetite would **find** it almost **an impossibility to** remain at table **with them.** The practice of throwing the elbows out, smacking the lips, making **a loud noise** in drinking or eating soup, belching wind, **clearing the** throat, blowing **the** nose, speaking **while the mouth is** full of **food, are in nearly all cases** done unconsciously from **the habits of childhood, but these things are** nevertheless **a fruitful source of annoyance to persons** who have been **more carefully bred,** and quite enough to destroy the **pleasure of a meal,** and in **a corresponding** degree detrimentally **affect digestion.**

EAT, DRINK, AND BE MERRY.

Upon no consideration **eat** heartily **while** suffering from mental or bodily **fatigue, or** during **a fit of** anger, or great mental **excitement of any sort. Have**

your dining rooms warm, light and cheerful; your table laid with scrupulous care and neatness; banish from the table every topic of an unpleasant nature. If family jars must take place, never allow the dining-room to be the battle-ground. It is not only an injustice, but an outrage upon a child to select the meal time, in the presence of the whole family, for correcting his fault. The mortification and anger must seriously interfere with digestion and nutrition, to say nothing of the depression of the spirit and the permanent influence for ill upon the mental and spiritual nature; and what is true of children is equally true of adults.

The individual who habitually goes to the table with his thoughts full of all uncharitableness and fault-finding, is unfit to sit at meat with the more harmonious and peacefully inclined. He comes as an element of discord where only good-will should reign. It is meet that a benediction should settle down over those at table; not merely a wordy one, but that the mantle of charity and peace should cover and warm all during these most important and pleasant reunions of the day, during which no disputing or bickering should be allowed. Light and agreeable conversation should take the place of weightier matters. Take time to taste your food and see how delicious it is, as the gusto assists in its digestion.

Thackeray, upon his visit to this country, was, in common with all other noted men of letters, fêted and dined to his heart's content. Once, while dining at Delmonico's, he was greatly annoyed by the incessant conversation kept up by the guests present, who evidently desired to air their wit and wisdom for the benefit of the humorous Englishman. He bore it for a time, but as one course after another came in, each one more delicious than its predecessor, he would break out with—"Now, boys, keep still while we taste this;" and would fall to eating with such keen relish that for a time he was seemingly oblivious to his companions, but between the courses he was as ready with his flow of wit and repartee as any one present.

Encourage the habit of sitting long at table, even after the meal has been finished; the pleasant conversation is a good digester, always bearing in mind that eating and digesting are the most important of all of the processes, and should have both time and attention bestowed upon them. The great prevalence of dyspepsia among business and professional men is largely due to the habit of hastily bolting their food and hurrying to their offices and places of business and at once entering upon their duties, becoming subject to all the harassment attending their varied avocations.

Of all the outgrowths of the passions, none so seriously affects the health of the digestive organs as does fretting and worriment of mind. This is especially noticeable in the case of women, whether it is because they have more to harass and annoy them, giving greater cause for those distressing manifestations of temper, or because they lack the stimulus of those larger, more ennobling and entertaining enterprises in which men are constantly engaging to counterbalance the little ills and crosses of life.

To sum the matter up in a few words—don't fret! Take life easier. Preserve your equanimity. Take your after-dinner siesta. Let your food be only of the best quality and prepared in the best manner. Eat only when hungry. In summer, incline to and enjoy the productions of the vegetable world. In winter, revel in your buckwheat cakes and syrup, roast beef, beef-steak, poultry, game, and all good things that are kindly toward you in their ministry. Secure your medicinal agents in your food supplies, for somewhere in the great laboratory exists an antidote for every ill of the flesh, which the careful student may with ease discover.

CHAPTER VI.

NERVOUS SYSTEM.

Lewes says, "In the mighty web of things there are no threads more wonderful than Sensation and Thought; nor have any more constantly solicited the attention of philosophers, from the earliest dawn of speculative inquiry to the angry contests of to-day. They have been problems ever alluring and ever baffling; one moment the threads seemed to be within the grasp of an outstretched hand, only to vanish again into the inextricable confusion of tangled mystery."

While we will not attempt to explain the real origin of Thought and Sensation, we can still give an outline knowledge of the anatomy, physiology, and hygiene of that most complex apparatus, the brain and nervous system. A knowledge sufficient to enable the reader to guard against many of the abuses growing out of our advanced civilization.

We are now somewhat familiar with the several systems in the physical economy, the circulatory, glandular, etc. We have still to consider the most important one, that which infuses and gives life to the entire structure. We have seen in the study of circulation of the blood that that fluid has a great common center to which it is attracted and from which repelled. This center is the heart and lungs, and in order that this fluid shall permeate the most minute portion of the organism there is a marvelous system of division and coalition of the arteries and veins, until they become so small that the naked eye can no longer detect their presence. No portion of the structure is so remote from the center that it does not receive thousands of these tiny rootlets.

The nervous system, like that of the circulatory, has a grand center, the interior of the brain, from which radiate innumerable white cords, delicate cylindrical filaments, called the white matter of the nervous system. These cords, like the arteries, throw off from the larger trunks millions of glistening threads, which dividing as they reach the surface of the body, present an array far outnumbering the combined armies of the world; each nerve fibre constituting a sentinel on the outer walls to guard the citadel within. In this system resides all sensation and motion.

There are two forms of nerve tissues—the white filaments described, and the gray matter, composed of grayish-red, or ash-colored cells of various sizes, possessing one or more branches of the white fibres; the gray cellular substance forms those important groups known as the nerve ganglia, to which the fibres combine, serving to connect them and placing them in communication with other parts of the body.

The brain is the largest and most complex of the nerve centers. It weighs in the white adult about fifty ounces, and consumes about one-fifth of the entire blood. The tissue of the brain is soft and easily injured by pressure, and is therefore protected by the skull and three membranes, the *arachnoid, dura mater,* and *pia mater;* the two former are tough, elastic coverings, the latter a delicate, web-like membrane which furnishes the brain with arterial blood. This coat dips into the convolutions of the brain and serves to keep the soft, pulpy mass in place. The brain cells are oval, oblong bodies, larger at one end and tapering off at the other, to each of which are attached a minute white nerve filament, the larger end lying outward toward the *pia mater,* the small end, with its nerve attachment, turning inward toward that central portion of the brain known as the *sensorium,* presenting to the naked eye a simple mass of white matter, while the outer portion of the brain shows only the gray cells.

When the surface of the organ is examined there are found deep furrows or convolutions; the extent of its area, if the wrinkles were smoothed out, would exceed, in a well-developed brain, four square feet, and yet in order to economize space it is gathered up in folds so that it may be contained within the small limits of the skull. The process of perfect intellection is due more to the depth of the convolutions and firmness of the brain texture than to the size.

In very young children, idiots, and the undeveloped races, the convolutions of the brain are very shallow. It is subject to the same law which governs the development of the muscular tissues, and is improved, strengthened, and developed by judicious mental exercise, gaining weight and density through years of active intellectual labor and thought. There is no doubt but what Webster's brain gained in weight as he advanced in years and in the accumulation of wisdom, just as a kernel of wheat grows firm and heavy as the life of the stalk is being crystallized in its depths.

Underneath the brain the myriads of nerve fibres meet together and form a large bulb-shaped mass known as the *medulla oblongata*, which, decreasing in size, passes out at the base of the skull, enters the vertebral column and forms the spinal cord. Its composition is like that of the brain, but the manner of its arrangement exactly opposite. We found the gray matter of the brain placed on the outside, and the white on the inside; in the spinal cord the white is on the outside and the gray in the center. The three membranes which enclose the brain also descend and protect the cord, which is separated into two halves by a lateral fissure running the length of the column. Each half is made up of two distinctly different bundles of fibres, and have separate works to perform. Between each bony ring composing the vertebra there is an oval opening on either side, through each of which pass two nerve fibres, one from the back and one from the front. Those from the back portion of the cord are distinguished from those of the front by being connected with a ganglia of gray matter. Those sent from the front are the nerves of motion, and from the back, of sensation.

We find that the gray cells and white matter perform different missions, the gray being active and originating nervous impulses, while the white filaments serve as conductors of the impulse, by which means the ganglionic system is enabled to communicate with the near as well as the remote portions of the body.

For the sake of illustration, we may consider the brain as the seat of government; while the gray matter in the spinal cord and the ganglia, like subordinate official posts, are in command of the out-lying provinces, the white nerve fibres acting as means of com-

19

munication between the provinces and the local and central governments. Just behind the stomach there is a complicated net-work of nerve fibres connecting a great number of small ganglia, and serving as a relay station for the surrounding organs, and doubtless controlling digestion and nutrition as the brain does intellection. Passing downward from this nerve center, and following the course of the main artery until they reach the pelvic cavity, are a vast number of nerve cords. Here again they weave themselves into a network, interspersed with tiny bodies containing the gray matter. This flexus lies in the pelvis, directly behind the genital organs, and evidently has the office, in the female, of supervision over ovulation and gestation, and the general health of the various organs engaged in this work. These nerve centers are as easily injured by pressure as the brain; so that clothing worn tightly in the region of the stomach will necessarily crowd the organs back against the sensitive ganglia behind the stomach, and suspend normal action, just as pressure on, or ligaturing of the limb puts the foot "asleep." If this be long continued, permanent injury will be the result and various forms of indigestion follow. In case of prolapsus of the bowels, and misplacement of the uterus, there is a pressure on the lower flexus, producing similar indications of embarrassment and disease as would be found in concussion of the brain, pressure of the limbs or stomach, and this compressed condition of the last named group of nerves, gives us the various manifestations of "nervousness" so prevalent among our women.

In accepting the fact that the brain is the great central seat of government for the physical economy, we must readily foresee the injury resulting to the general system from any demoralization at headquarters. Two-thirds of the prevailing diseases have their origin in the disturbed brain, particularly the various disorders of the nerves.

Seguin, in his extended work on nervous diseases, speaks on the subject thus: "The growth of physiological and psychological knowledge in the past few years has caused mental affections to be classed with nervous diseases. * * It should be borne in mind that many nervous diseases, so-called, are only expressions of general pathological states, or sympathetic reactions to local morbid states of non-nervous organs. It has been thought that certain nervous diseases, such as insanity, hysteria, epilepsy, etc., become more frequent with increasing civilization. This is not fully established, and yet there can be no doubt that the strains of social life, the struggle for existence, the enormous striving of ambition, the intemperate use of sensual gratifications, cause the above diseases in a more or less direct manner. Nervous diseases—or, more exactly speaking, the liability **to** nervous diseases **—are** very easily transmitted from parents **to** their children, this being most strikingly shown in insanity, hysteria, epilepsy, neuralgia, apoplexy. **An** important factor in the development **of** nervous diseases is wrong education, the cultivation of the mental powers during the **age of** growth; not enough rest, **and** insufficient (especially fatty) food being allowed. The evil effects of school life are seen in **both sexes** though perhaps more often in the **female.**"

The brain is thrown out of **balance by the** unnatural pressure **of** continual **anxiety,** care, undue **and** prolonged excitement, fear, remorse, constant fretting, anger, jealousy, insufficient sleep,—conditions tending to produce irritation **of** the great center of the nervous system, and like a worm at the root of a tree, **soon** destroys both leaf and blossom.

Pathology must recognize this important psychological influence before it can hope to cope successfully with **disease.** The fact that the healing power comes from within instead of from without, is an all-important **one** which the patient **as** well **as** the medical adviser should **most** thoroughly comprehend. The influence exerted **at** times over **the bodily** functions by the imagination and will, is paramount to that produced by the most efficient of therapeutic agents, which at best can only assist nature in her work of recuperation. Mental disquiet produces physical discord in a greater or lesser degree. One hour of intense mental suffering reduces the bodily vigor more than days of severe physical and mental exercise performed under the stimulus of a contented mind.

The brain and its nerve attachments serve as a delicate instrument by which the soul manifests itself to the world of matter, the grosser parts of the body forming an envelope to encase that delicate and subtile apparatus from which the body derives its life, and which outwardly manifests the health and strength of the great central ganglia.

The brain does not produce thought any more than the hand does the fruit which it plucks; it simply serves as a medium of transfer of thought and emotion, as the hand does for the transfer of the fruit; and furthermore, is subject to the law of development and deterioration which governs the structure of the hand, which a normal exercise enlarges and strengthens. The musician understands that in order to acquire the greatest facility in instrumental execution, constant practice is necessary to keep the muscles of the arm and hand in a firm and flexible condition.

Dr. Winship, by a persistent course of physical training, arrived at a point where he could lift something
over two thousand pounds, while men of fully twice
his weight and in firm health could not raise five
hundred. The Doctor's prowess did not arise from
superiority in size, but from a development resulting
from a judicious muscular training. What is true
of muscular, is equally true of brain tissue. A given
amount of systematic and consecutive thought daily
is as essential to the health and longevity of a man or
woman as would be a reasonable exercise of the muscular system. Not thoughts in a downward direction,
which fret, worry, mar, and scar the soul; but in an
upward and outward course, leaving self out of the
question. The forgetting of self is often the losing
sight of bodily suffering and infirmities. One-fourth of
all the diseases which afflict humanity arise from an
intensified self-consciousness and self-love.

Men and women in magnifying their ills and wrongs,
in dwelling upon their real or supposed maladies, in
nursing every little ache or pain, or trifling wrong done
them by another, are constantly sowing the seeds of
disease. We are too liable to fall into the habit of
rehearsing over and over again all of the dark scenes of
life, and bemoaning our fate as if no one else ever had
a sorrow. We crawl into the shadow and sit and whine

about the darkness, as though God had purposely hidden his sunshine from us; when the truth is we are too indolent and selfish to get out of the shadow and seek the sunlight which is glowing in splendor all about us. The narrow, bigoted and censorious soul can do but little for its tabernacle of flesh. It is only in proportion as it becomes emancipated from selfishness, and emerges into the open field of thought, research and unselfish endeavor, that the body develops and grows strong to meet the increased needs of the enlarged soul.

It is just as evident that the cranial viscera has its divisions, each division having a separate office in the intellective process, as that the thoracic, abdominal and pelvic visceras contain numerous organs, each performing totally opposite functions. To clothe, shelter, and feed the body requires only a limited number of the faculties of the brain; the ordinary routine of daily life which provides for the physical needs alone, calls into play merely the basilar portion of the brain, as the processes of digestion, absorption, and assimilation involves the use of the abdominal region and not the thoracic and pelvic. The disposition to seek food, shelter, and warmth is prompted by an instinct possessed by all animals, man alone having a development of the brain evidently intended for superior intellection. A brain constantly engaged in the battle for earthly possessions must become one-sided in its development, just as the muscles would were they used only in one direction. The world is full of one-sided people who have turned all of their energies into one channel.

Men striving to amass wealth, gain notoriety or fame; women submerging their physical, mental and spiritual identity in the shallow tide of every-day life, in the cares of their homes or the demands of society and fashion, all of which, instead of serving as means of development, become a source of care, anxiety, over-work and unhappiness.

Long continued physical exertion, unrelieved by mental exercise and suitable recreation, will not only impair the health, but shorten life. Close mental application unencumbered with anxiety would be far less destructive to both brain and body. The literary men and women of the day furnish proof that a con-tinuous and judicious exercise of the brain is a means to physical preservation and longevity.

Those men and women who have lived continually in a world of thought and ideas and have developed their brain resources are just in their prime at the age of forty, providing they have obeyed the laws of health in their eating, drinking, and sleeping, and their best work is often performed near to, or after the fortieth year. Among our representative country women may be cited as examples, Mary Livermore, Elizabeth **Cady Stanton**, Harriet Beecher **Stowe**, **Julia Ward Howe**, **Mary Clemmer Ames**, **Mrs. J. C.** Croly (Jennie June), Mary Mapes Dodge. And **Mrs.** Trollope, the English novelist, did not commence **her** literary labors until past **her** fiftieth birth-day. These women **have** labored **on** without interruption **for years in the** enjoyment of health and mental vigor.

Wendell Phillips **is now** approaching his seventieth year, yet scarcely **looks fifty**; although he has been **for** over forty **years engaged in** that most **wearing of all** callings, a public lecturer, there appears **no abatement** in the power **of** his thought or force of his eloquence.

John B. Gough's head is whitening under **the accu**mulated snows **of sixty** winters, yet underneath **the snow** the fire of **his** genius **burns as brightly as it did** twenty years **ago**.

Henry **Ward** Beecher's intellectual powers were never greater **than at present.** Will Emerson ever grow too old to think as only **Emerson can think?** Our poets are nearly all **men of** ripe years, **and as** they advance in age are still **growing** richer in thought **and** readier in expression. Whittier, Longfellow, Tennyson, Buchanan, and Bulwer Lytton, are **all men far** past middle life.

Those persons **whose pursuits demand broad** mental scope usually live to an advanced age, and retain, to an astonishing degree, their intellectual and physical powers. Emma Willard lived to see her eighty-fourth year, and her life was one of great mental and physical activity and rich in good results. Mrs. Sommerville lived until within something less than a month of the ninety-second anniversary of her birth. Her last work, "Molecular and Microscopical Science," considered by far her best, was written when she was near her ninetieth year. Sarah Jane Hale approached near to the nineties when she closed her work in life, and for fifty years she wielded her pen with little or no intermission. Hannah More lived to see her eighty-eighth year. Elizabeth Montague held her mental faculties unimpaired up to her eighty-second year. Franklin had reached his eighty-fourth year, and to the hour of his death manifested the utmost activity and clearness of the intellectual faculties. Guizot, the French historian, was also past eighty-four when he died, and up to within a year of his death he prosecuted his literary work with unabated vigor, and walked several miles each day.

Michael Angelo painted his celebrated picture, "Last Judgment," when something past the age of seventy, but grew better in his art every succeeding year, up to his death, which occurred after his ninetieth birthday.

Titian, it is said, painted at the age of eighty-one that masterpiece of art, "Martyrdom of St. Lawrence," and it is authoritatively stated that for eighteen years afterwards he still worked on without a perceptible diminishing of his wonderful powers.

Madame Genlis celebrated her eighty-fourth birthday, and the greatest portion, as well as the best, of her works were composed and written during the last thirty years of her life.

Men and women who write, paint, carve, or in any way work for the good of humanity, grow strong in both brain and body, in proportion to their interest for the welfare of those for whom they labor. The philanthropist who gives his time and talent to the amelioration of his fellow-men forgets to grow old. The scientist who penetrates beneath the rough rind of the earth for its hidden treasures with which to better the conditions of men, loses sight of self and selfish aims in contemplating the magnitude of the earth's resources and the wonders of science which he is to interpret to the world, and the strongly-energized brain vitalizes the body, which seems to defy decay. Baron Von Humboldt gave to the scientific world the treasures of over half a century of study and research. At eighty he was still vigorous and at work.

Who can think of Huxley, Tyndall, Spencer, Carlyle, or Victor Hugo as growing old?

Agassiz was at his best when they celebrated his fiftieth anniversary; and it was *accident,* not natural decay, that robbed the world of the benefit of his labors.

The projectors of commercial and financial enterprises, which have had for their basis the general good of mankind, have most frequently been long-lived, showing to the last a remarkable preservation of the faculties. Few mornings are too stormy to prevent Peter Cooper—now more than eighty years of age—from attending to his business. Commodore Vanderbilt furnished an example of long-continued mental and physical labor, accompanied by a wonderful intellectual and bodily vigor.

All merely selfish aims retard the development of the brain, and as the sphere of life narrows and the aspirations become less elevating, it loses its hold upon the body, which falls to disuse and decay.

What man needs is an equal use of all his powers. To cease to use a faculty is to lose it; it matters not whether it be of body or mind. He needs work first, recreation next, and rest last. King Alfred recommended that the day should be divided into three divisions: eight hours for work, eight for recreation, and eight for sleep. The recreation should differ according to the nature of the employment. Those exercising the brain exclusively should choose a kind of recreation which tends to call the blood to the muscles and thus rest the brain, such as walking, riding, dancing, driving, base-ball, archery, boating, skating, etc. ; and those who use the muscles entirely should seek that sort of recreation which would cultivate the mind and rest the muscles, by drawing the blood to the brain, as that fluid flows to those parts which are most active. Rest does not mean suspension of action, but a change of occupation.

The disease known as softening of the brain rarely afflicts those persons who have properly and vigorously used that organ. It is no more an indication of overwork than a pale, flabby, bloodless muscle would be. Brain-softening is more frequently the result of long-continued feverish excitement, mental harrassment, dissipation, an excess of physical exercise, or sexual abuse ; or, on the other hand, idleness, luxurious living, and inactivity of the brain, but never from appropriate and systematic mental labor.

Where the occupation is in every way distasteful, producing a feeling of dissatisfaction and weariness, it will most surely in time bring on both mental and physical disturbances, and if within the range of human possibilities should be changed for something better suited to the tastes and capacities of the individual.

The world is full of work, and of so varied a nature as to meet the requirements of all. The principal cause of the nervous suffering, unhappiness, and half-dead-and-alive uselessness of one-half of the people in this world, is that they, by mistake, get into the wrong place, and can see no way of escape; therefore, as the law of fitness is inexorable in its rule, they are continually under the lash while so placed.

A man or woman compelled by force of circumstances to perform manual labor wholly, when their intellectual endowments would more perfectly fit them for mental work, will suffer constantly in mind and body. The same is true of those whom circumstances have placed in positions for which mentally they are incapacitated.

Parents and guardians should understand this law of fitness, in order to place the children under their charge in positions best suited to their capacities, regardless of sex, or position in society. Many a good artisan, mechanic, or farmer has been spoiled in a doctor, lawyer, or minister, and *vice versa.* There are comparatively few who would not make a success of life if they were started in the right direction.

For instance, take the cases of scores of our young girls who possess no talent for music; let one-half of the time and money now expended upon that study alone, be applied to the development of their more available talents, with the privilege of using those talents, and they would become useful, happy and healthy, instead of fretful, dissatisfied and nervously diseased women.

> " There is a tide in the affairs of man
> Which, if taken at the flood, leads on to fortune;
> Omitted, all the voyage of their lives
> Is bound in miseries and in shallows,
> And they must take the current as it serves
> Or lose their venture."

The important object in life is to seek the sphere in which we can work most acceptably to ourselves and efficiently to others; for work we must, in one direction or another, if we would develop the best portions of our nature and be healthy and happy.

CHAPTER VII.

ANATOMY AND PHYSIOLOGY OF THE FEMALE GENITAL ORGANS.

As WE carefully survey the arrangement of the human body we notice that every organ or set of organs has its appropriate place in which to perform its several functions. We likewise notice that pressure, however slight, will produce, sooner or later, functional derangement of the parts. For example, if the skull be fractured and pressed in upon the brain the sixteenth fraction of an inch consciousness will be suspended until the pressure is removed; therefore we find all of the more vital parts surrounded and protected from injury. The brain by the skull, the spinal marrow by its bony rings, the heart and lungs by the breast-bone, shoulder-blades, ribs. Below the abdominal viscera, we come to the pelvis, a bony basin, in adult life composed of four bones, the innominata, two bones forming the side and front walls, more familiarly known as hips and pubic bones, and the sacrum and coccygis at the back, uniting it with the spinal column. This bony cup contains, in the female, the generative organs, the arrangements for the protection of which must be obvious to every observer. If an erect position of the spinal column is preserved there will be in the lumbar region, the small of the back, a deep inward curve, which will draw the pelvis backward and throw the organs in the abdominal viscera outward, preventing them from settling down upon the group in the lower cavity.

This matter of position is of the first importance, as no real advance can be made toward a permanent cure until all organs have attained a normal position in the body.

A brief survey of the anatomical structure of the contents of the pelvis will be all that will be necessary for our present purpose. The group is composed of the uterus or womb, ovaries, fallopian tubes, broad and round ligaments, and vagina. The uterus is a flat, hollow, pear-shaped muscular organ, varying in length from two and a half to three inches, by one and a half to two inches broad at its widest part; the neck, in health, not exceeding an inch in width; the walls are composed of firm, elastic bands of muscles, so arranged as to get the greatest leverage during parturition. We find also that this structure is eminently vascular. The blood veins and arteries are connected with the great aorta and vena cava by two veins and arteries on each

21

side, entering the organ near its mouth, passing upward
and inward, forming a system of division and subdi-
vision, ending in minute capillaries, presenting their
open mouths to the interior lining; and through these
not only is the menstrual fluid expelled, but during gestation the fœtus receives its nourishment in form of
arterial blood. When free from disease and unimpreg-
nated, these vessels in the uterus are very minute, in
fact merely rudimentary; but as soon as conception
takes place or some disturbing condition like misplacement arises, at once all of the veins and arteries
become engorged and congested. In conception the
blood is absorbed and consumed by the embryo as
fast as received, but in the case of any abnormal
condition the influx of blood becomes a fruitful source
of irritation and disease.

The uterus hangs in the pelvic cavity like an inverted
pear, the neck attached to the walls of the vagina, a
muscular tube leading from the mouth of the uterus
to the external opening in the body. Attached to the
fundus or upper part of the uterus we find the ovaries
and fallopian tubes. The ovaries are connected by two
round firm ligaments with the body of the womb, into
which the fallopian tubes open, their office being to
carry the prepared ovum from the ovarium to that
organ, the transfer being accomplished through the ver-

micular movement of the muscles of the tubes, after the manner in which food is propelled through the intestines, the ovaries having no other means of communication with the uterus. Just below the two previously named appendages we find the round ligaments; these are firm, elastic cords, which pass from the uterus outward and attach themselves to the pubic bone for the purpose of holding the organ in its place, at the same time allowing it to expand and rise, as during pregnancy. Enclosing these three appendages there is a tough membrane, known as the broad ligaments, which unites with the posterior ligaments attached to the lower portion of the vertebra. This broad band serves the purpose of holding the group in the center of the pelvis as well as affording a means of protection. These organs are further guarded anteriorly by the bladder and posteriorly by the rectum.

The vagina is about four inches in length, and lined with a mucous lining arranged in folds or wrinkles in order to accommodate itself to the expansion of the outer muscles during childbirth. This organ in a healthy condition acts as a support to the uterus, *prolapsus uteri* often arising from a loss of tone in the walls, which settle down and drag the organ with them. This relaxed condition can be greatly relieved by a series of voluntary contractions of the muscles of the

lower portion of the vagina. These movements should
be performed at regular intervals and in connection
with contraction of the abdominal muscles, a horizontal position being the most convenient for the
purpose. This effort of the voluntary muscles performs
for their structure precisely what rowing or boxing
would do for the extensors and flexors, always providing that the former are as free from pressure as the
latter.

MENSTRUATION.

After the menses are fully established, in healthy
females there will be a regular recurrence of the flow
every twenty-eight days or thereabouts, varying slightly
in different individuals. This periodical flow continues
until the change of life, except during the periods of
pregnancy and nursing, when, as a rule, it disappears.
The menses have a two-fold mission evidently, one to
prepare the uterus and genital organs for the recep-
tion of the ovum or egg, the other to cleanse the
general system preparatory to the evolvement of a new
being, for the ovum is the female germ of generation;
and every month this preparation and ovulation takes
place in single as in married ladies, from the period of
puberty until the change of life.

Menstruation is one of the important processes of

life, the same as thinking or digesting; there is something evolved from the process, and this evolvement requires vital force. It therefore follows that during the menstrual period, as in cerebration and digestion, that there should be an entire freedom from mental and physical excitement, during the earlier portion especially; indeed, absolute rest is imperative where the physical well-being is to be taken into consideration. An over exercise of the mental faculties coupled with anxiety of mind during the menses is usually the cause of amenorrhea. Particularly is this true in the cases of young girls who are attending school at a distance from home and afflicted with home sickness. This suppression, if not causing death, often results in serious disorders of the genital organs; hardening of the uterus, profuse flooding, leucorrhea, etc. These conditions are induced just as brain difficulty is, from attempting to do too many things at the same time, or as dyspepsia is caused, from using both the brain and the muscles while the stomach is engaged in digestion. During the first two days, especially, of this process, there should be quiet and rest, as far as it is possible in the nature of a woman's duties. This temporary suspension of activity will be more than recompensed by the feelings of relief and buoyancy which will result from it. The comparative freedom of the Indian women from suffering during parturition is attributed to the fact that they are permitted the utmost quiet during the menses. They are not forced except under extreme circumstances to march with the tribes, but are accorded all of the leniency and care due to invalids.

When the attention has been aroused to the importance of this subject it will be found to be comparatively easy to adjust matters, so that quiet during this period may be observed, in which case a greater attention can be paid to cleanliness by frequent local bathing with warm water; then, to a degree, the pernicious habit of using napkins could be dispensed with, as these articles collect and hold the menstrual fluid near that portion of the person the nature of which is to absorb whatever comes in contact with it. The discharge is often acrid, and at times poisonous, and should be at once expelled from the body, which is impossible under the circumstances; the cleansing process is never so perfect, inflammation and granulation of the vagina is often induced, and after a time, inflammation and ulceration of the mouth of the uterus. This unwholesome practice taken in conjunction with the already engorged and hypertrophied condition of the misplaced organs, and we can readily comprehend how complicated the case would inevitably become. At the cessation of the flow, hot injections should be used in all cases, as the fluid will collect in the vaginal folds and become to a degree putrescent. A small amount of borax may be added to the wash as it cleanses and sweetens the mucous coat.

There is no obstacle in the way of any young girl taking an injection after the menses have been established; as the hymen, unless abnormal in its formation, does not interfere with the passage of a small tube, such as the one attached to the Fountain syringe; nothing larger should be used.

There is still in existence an old superstition that there should be no water used on the person during the menstrual period. There could not be a greater fallacy. We would advise an abstinence from general cold baths, but local baths of warm or hot water, as the case may require, are really indispensable to comfort and good health. For painful menstruation, hot hip baths, as hot as the patient can bear, should be administered, and in most cases hot injections; and when the Turkish bath can be obtained, it should be resorted to, as it is of the greatest value in all cases of obstruction. Hot vaginal douches are at all times beneficial; not merely warm ones, but as hot as the parts can bear. They can be taken at the temperature of from one hundred and ten to one hundred and twenty degrees Fahr., always making use of the Fountain syringe, and using from two to four quarts of water at a time. This is a fine stimulus to the vagina when it has lost its tone; it also stimulates the uterus and assists that organ in cleansing the overcharged walls; and when one is greatly fatigued from being upon the feet for a long time, a hot injection will soothe and give relief as nothing else will. For great weakness of the parts, it is of the utmost value. It must be always borne in mind that *heat* is positive in its action; tepid or merely warm applications are weakening or rather sedative in their effect, and should not be used except in excessive inflammation or in cleansing of wounds, injections after child-birth, etc., where a soothing effect is desired.

Cold vaginal injections are only admissible in extreme cases of uterine hemorrhage, but ordinarily the shock will produce chronic congestion, and induration of the uterine walls. This result is almost universal where cold injections have been resorted to as a prevention of conception. It would be difficult to place an estimate upon the injury to the brain alone, to say nothing of that done the sensitive genital organs, from this pernicious habit.

In case of profuse menses, the hot injections taken between the periods often correct that difficulty; but in this, as in all other cases of derangement of the female organs, there must be absolute rest at the menstrual period; and after that has passed and the parts have recovered their natural tone, introduce into the vagina every alternate day for ten days, a ball of cotton saturated with glycerine, to which has been added tannic acid in the following proportion: Glycerine, five parts; tannic acid, one part; allowing the cotton to remain twenty-four hours, unless there exists leucorrhea, in which case only allowing it to remain twelve hours, using a hot injection after removing it. Tie a bit of cord through the cotton by which to remove it from the vagina.

There is no regular standard as to the length of the menstrual period, or the quantity of fluid lost. Each woman is a law unto herself, and that which would appear to be a hemorrhage in one case proves only to be the normal condition of the individual. Much depends upon the rapidity with which the system manufactures blood, as well as upon the amount consumed in the various processes of life. It may be more rapidly used up by the tissues; the absorbents may be more active; the brain may be called upon to do a greater amount of work in one case than in another, in which instance a larger quantity of blood would be consumed; consequently there would be less to lose in the menses. When the individual is of full habit, accustomed to high living and luxurious ease, taking only limited exercise of brain or body, a profuse monthly flow is a safeguard against repletion of the system. In many cases the loss is very slight, yet the persons are perfectly comfortable and healthy.

That the flow should be kept uniform as to time and quantity, is the important consideration. The treatment should be in all cases to encourage an increase, rather than a decrease, in the quantity of the menstrual fluid while approaching the change of life; it is an effort of Nature to reduce the system somewhat, in order to prevent plethora and its attendant evils, when the drainage shall cease altogether. Those who have passed the change with least injury to brain and body are those who have for four or five years previous had a very profuse flow, amounting almost to a hemorrhage each time.

A gradual lessening of the quantity before the proper time for the change of life should be regarded with distrust, and every available means used to establish a normal flow, such as hot foot and sitz baths, hot vaginal douches, Turkish baths, much exercise on foot, all tending to attract the blood away from the overcharged vital organs to the lower extremities. There must also be a lowering in the nutritious quality, as well as a lessening in the quantity of the diet. Repletion of the system is much more to be dreaded at this important period of life than depletion. In the former case heart, brain, and nervous difficulties are almost sure to manifest themselves in more or less aggravated forms, to say nothing of the danger of cancerous and tumorous formations upon the ovaries and uterus.

A gradual decrease in the menses, dating from the twenty-eighth to the thirty-fifth year, is more frequently due to induration of the uterus than from any other cause; a condition which utterly incapacitates that organ for the work of drainage, which in a normal condition it evidently performs. Each month the menses are carried to the uterine walls to find an outlet through the mucous lining. If all of the avenues in form of blood veins are already blocked up with the blood which was not properly drained off on the previous month, then this new influx must flood back and be partially taken up by the absorbents and poison the circulation; but after a time Nature accommodates herself to this condition of things, and in order to prevent death from the accumulation of fluid in the already overloaded parts, sends the surplus heat off in other directions, and in time we have an increased deposit of adipose, producing fatty degeneracy of the heart, an undue accumulation about the vitals and abdomen at times, a growth of fatty tumors in various portions of the body.

The hardening of the uterus arises from a variety of causes; but most frequently from long-continued exposure to severe cold during the menstrual period, while the blood vessels in the uterine lining are lacerated; the use of cold douches and other abnormal means to prevent conception, using mechanical contrivances for the purpose of bringing about miscarriages, accidental miscarriages, strangulation from the prolapsed bowels and the more aggravated forms of misplacement, such as anteflexion and retroflexion, conditions tending to prevent the escape of the menstrual flow; the wearing of napkins during the period until they become acrid and even putrescent; the poison communicating itself to the linings, thus producing congestion and ultimate hardening of the walls of the uterus, which become turgid, like a hardened and inflamed tumor. We find scores of women of twenty-five and thirty where the menses have almost disappeared and the body of the uterus in the condition above described, which does not arise from an isolated wrong, but from a series of transgressions extending over years. The recuperative powers are at all times making strenuous efforts to maintain an equilibrium, and it is only when overcome by great odds that they yield the point. Little by little, imperceptibly to the individual, these repeated offenses are undermining the vital forces, and **suddenly** there comes **a** great crash—Nature succumbs, **and** Therapeutics stand powerless, bewildered, and **to a** degree **useless.**

DRESS AND ITS RELATION TO FEMALE DISEASES.

Pathology demonstrates **the fact that** during the past **fifteen** years that class **of disease peculiar to** females **has** been steadily **on the increase; and very** naturally **the supposition arises that** there **must be some** general **cause** developing **this** tendency **towards** functional and organic derangement **of** the genital **organs, and the** verdict is almost **universal among** those **physicians who make a** specialty **of these** difficulties, **that** they **are** largely **the** result of the improper mode **of dress adopted by our women.** First, **from its being too tight or so** inconveniently arranged **as to prevent the free action of the internal** organs; second, **from the great** number **of bands with heavy skirts, resting** entirely upon **the** delicate walls **of the abdomen, causing the** intestines to fall down **upon the organs in the pelvic cavity,** producing strangulation **of the uterus, ovaries,** fallopian tubes—in fact, of **all of** the organs in the viscera. **From** this continued pressure **we find congestion** of the **uterine** walls, **and from the** long-continued **heat a demoralization of** the internal linings, **often** inducing that much-to-be-dreaded disease, dysmenorrhœa. There is also an involvement of the ovaries and fallopian tubes, a chronic congestion induced, leaving them susceptible to abnormal growths, and the individual to every form of abnormal menstruation, from the most profuse and painful down to a premature stoppage of the menstrual flow, by which Nature is baffled in her cleansing process, and as a result there are set up independent growths in the form of uterine tumor, both cystic and fibroid, and the various types of polypi, as well as the several cancerous formations; and descending in the grade, we have ulceration of the *os uteri*, leucorrhœa, catarrh, and a long catalogue of the lesser evils.

There is no good reason why menstruation should not be as painless a process as that of thinking, digesting, or of respiration.

Pain is the voice of Nature calling for a cessation of hostilities, unmistakably telling us that the law of discord has been substituted for the law of harmony. The same law which governs respiration and circulation also controls the genital organs. If from accident or an improper mode of sitting or dressing, the chest becomes compressed, and the heart and lungs crowded, preventing a free passage of the blood to and from the cavities, there will follow a more or less serious functional, and after a time organic derangement, in form of hypertrophy of the heart, ulceration or hepatization of the lungs; but give these organs fair play, and plenty of pure air, and they will take care of themselves; and in a like manner the female organs of generation will uncomplainingly do their allotted work without giving the amount of suffering we are continually called upon to witness.

Owing to the flexible nature of the abdominal walls, no weighty clothing should be permitted to rest upon the hips, but should instead be supported from the shoulders entirely. No bands, not even upon the drawers are admissible, for if the clothing is arranged on bands they will be constantly settling down while standing or walking; and in the act of sitting down, the strain on the clothing at the back will cause the bands to cut in across the front portion of the body where there are no bones to resist the pressure. This continued strain, together with the weight of the heavily trimmed skirts, will in *all cases* produce prolapsus of the bowels and the inevitable sequence, falling of the uterus.

We have seen that the trunk of the human body is divided into four parts: the thoracic cavity, containing heart and lungs, pericardium, and pleura; in the hepatic region, the liver, stomach, pancreas, and duodenum, this group lying just under the floating ribs; in the abdominal viscera are the small and large intes-

tines, lacteals, mesenteric glands, kidneys, and various
tissues; in the pelvic cavity, the bladder, rectum, uterus,
fallopian tubes, **broad** and **round** ligaments, and their
enfolding membranes; **all of** these, **to say** nothing **of**
the **millions of** blood veins and **arteries,** nerves and
nerve ganglia and lymphatic glands which ramify and
group themselves **throughout the** entire trunk. **We**
perceive that there **is** no surplus space, that each **com-
partment is full.** The lessening **of** the size **of any** one
of these cavities presupposes **one of** the two sequences:
either **that** the organs occupying the cavity are com-
pressed and strangulated, **or that they** are forced **into**
the sphere of another **organ or set of** organs, the results
being **virtually the same.** **Compressing** the **lower por-**
tion **of the waist where the** floating ribs **make no resistance, would, as a matter of course, force the** hepatic
group downward into **the** space intended for the intestines, and **they, in turn,** sinking still lower until they
encroach **upon the domains of** their neighbors, **the**
pelvic group; developing from time to time pathological
changes more **or less** alarming in **their** nature, **for the**
difficulties become after **a time not special but** general.
The American women, **as a rule, are** delicately organized. There **is a** lack **of** development in both the bony
and muscular structures, the muscular tissues have no
resisting power, the abdominal **walls are** weak **and**

flabby, and afford little support to the heavy internal organs, and this is why tight and heavy clothing is so especially destructive in their cases. Not more than one out of ten of all of our young women to-day has a perfectly formed pelvis. The unnatural manner in which they are trained prevents a perfect physical development. At the period of puberty, when the pelvis should broaden to accommodate the increased size of the genital organs, the weight of clothing, together with a lack of physical exercise, thwarts Nature, and lays the foundation for untold suffering at the menstrual period, and still greater agony, and perhaps death, during parturition.

CARE OF YOUNG GIRLS.

The majority of women are nearly every hour of their lives transgressing some one of the laws of their being. They do it ignorantly and thoughtlessly, it is true, but nevertheless the penalty is as rigorously enforced as though they sinned knowingly. Little by little they learn through harsh experience those lessons which should have been taught them in early girlhood, from year to year, as fast as their needs demanded the knowledge. It is a huge mistake to keep girls ignorant of the laws controlling their physical being. Knowledge would be a safeguard against temptations fre-

23

quently thrown in the way of young girls during that period of their physical development when they would be more **likely to** be influenced—through the nervous excitation attending the establishment of puberty—in the direction of wrong. Every mother whose young daughter has gone astray will some day, if she has not already, awake to the consciousness that she was largely responsible for the wrong.

During the period of preparation, and until puberty has been fully established and the organs relieved, through the **flow, of** the heat and excitement always attending this change, there should be the utmost care exercised over the body **and mind of** the young girl. There **is often a** degree of pruriency **existing in** both boys and girls at this period. **It is** no more a sign of depravity than an inflamed brain, throat, or stomach would be. It is simply a physical condition, and not an indication of moral degradation, and requires soothing remedies, just as any other inflammatory **state** would. Regulate the diet carefully, taking out all articles of a stimulating nature, giving only that which is substantial and nutritious. Let the course of reading be such as appeals to the **reason** and the higher senti**ments.** Prevent the reading of exciting love stories, also the promiscuous mingling of the sexes without the presence of older and wiser companions, who would check thoughtless indiscretions and who would turn the thoughts into different channels. Discourage all tendency toward morbid desire to be alone and to the indulgence in reveries. Give the young people plenty of physical exercise. Encourage them to use up the surplus heat of the system in physical exertion, as exercise calls away the excess of heat from the over-charged parts.

The sexual idea will continually intrude itself upon the young mind at this period, just as hunger would when the vascular coat of the stomach was engorged with blood. There is a constant prompting of such thoughts because there is a constant irritation of the genital organs, which in due time will subside and a normal condition be established.

This is a trying time in the life of a girl or boy, the period when a father, mother, or guardian has a greater responsibility imposed upon them than at any other. It is the turning point which makes or mars the character, as the generative organs have an important bearing upon its formation. Love is the pivot upon which the world moves, and for a normal and healthful manifestation of this passion there is required a perfect and normal development of the physical structure.

This truth has been fully demonstrated in the cases

of malformation, castration, and arrested development;
the castrato speedily losing all love and friendship for
friends, lover, and relative, as in the case of Abelard.
The same traits of character manifest themselves in
females who have been born without the ovaries, or who
have lost them through ovariotomy or spaying. As a
rule, up to within a year of the commencement of
puberty there is little or no life in the organs of generation; they are almost rudimentary. The exceptions
to this rule would be the children of licentious and
sexually diseased parents. Such cases are of course
numerous, giving rise to the monstrosities in youthful
crime which so shock from time to time the reading
public.

Puberty is clearly a period of development of those
most important organs. Any excitement or association
of ideas that would lead to a premature indulgence of
the passions must prevent a normal growth of the same.
Everything, therefore, that would, in any way, tend to
excite the sexual nature of a child should be shunned
as one would shun a deadly contagion. All allusions to
love, marriage, or flirtation in connection with very
young people and children are not only ridiculous, but
to a degree immoral. The practice in which some
people indulge of constantly joking children about
their "little" husbands, wives and sweethearts should

never, *on any account*, be permitted. A manifestation
of sexual precocity is almost always fraught with danger in the cases of both **sex.** The young mind should
be rather led away from this line of thought than
toward it, as it leads to pruriency, and *that* to nearly all
the wrongs and indiscretions of youth.

It also leads to early marriages, than which there is
but one greater misfortune to a young man or woman,
and that is, being the offspring of such a union.

But perhaps the greatest evil arising from this forced,
unnatural growth of the sexual nature is that of self-
abuse, which, owing to the premature development of
the brain and nervous system of our children, is daily
on the increase and much more prevalent than is
generally supposed. Dr. Adam Clark says: "In my
opinion neither the plague, nor war, nor small-pox, nor
similar diseases, have produced results so disastrous to
humanity as the pernicious habit of onanism. It is the
destroying element of civilized societies which is constantly in action and gradually undermines the health
of a nation."

Dr. Gardiner, formerly of Bellevue Hospital, says:
"Much of the worthlessness, lassitude, and physical and
mental feebleness attributable to the modern woman
are to be ascribed to these habits as their initial cause.
Foreigners are especially struck with this fact as the

cause of much of the physical disease of our young
women. They recognize it in the physique, in the
sodden, colorless countenance, the lack-lustre eye, in
the dreamy indolence, the general carriage, the constant
demeanor indicative of distrust, mingled boldness and
timidity, and a series of anomalous combinations which
mark this germ of physical and moral decay."

From this outrage on Nature may be expected almost
any form of sexual abuse and disease; manifesting
themselves variously in the sexes, prominent among
which may be mentioned in the case of boys—seminal
losses, epilepsy, and gradual imbecility; in girls—
hysteria, nymphomania, acute inflammation of the
female organs, and in common with boys, epilepsy and
mental derangement. The irritation, leading to these
disastrous results in females, is often produced through
want of cleanliness, acrid discharges from the uterus
and vagina, from walking long distances in hot weather
and omitting to cleanse thoroughly with soap and water
afterwards. This inflammatory condition is frequently
caused in very young female children by what is known
as pin worms creeping into the private parts from the
rectum. Free daily bathing of the external genital
organs is absolutely imperative in all cases, that they
may escape the serious results of irritation in these
parts.

Not an ounce of clothing should be permitted to rest upon the hips and abdomen of a young girl while approaching or passing through that most important change of life. She should be permitted the utmost freedom to exercise physically in all directions, walking, riding, romping, if she is so inclined; but if she has a morbid tendency and is averse to exercise, then it is the duty of those having charge of her to encourage her to out-door exertion, and plenty of it too; always carefully guarding her body from compression and heavy clothing that there may be a natural free play of all of the internal organs.

Let the exercises be taken between the periods, but strictly observing quiet and freedom from excitement during the process of ovulation. Sitting with cold, damp feet at the time or just before or just after the flow must never be permitted, as the system is more susceptible to sudden shocks during the early stages of this period than later in life, although such exposure would be hazardous at all times. The body should be warmly clothed, and if in a cold season of the year and the menses painful, a perfectly-fitting, broad, double-flannel compress must be worn over the abdomen to prevent a chilling of the surface. If the child has naturally any physical stamina and no deformities of the vagina, then if these precautions are observed there is no danger but what Nature will take care of the rest of the process.

HOME TREATMENT OF **FEMALE** DISEASES.

Gartering the limbs, wearing the shoes too tightly **buttoned** about the ankle, attaching the stocking-supporters **to** a band about the waist, as **well as** allowing the clothing to rest upon the hips, form **a system** of ligaturing which must produce an overcharging **of** all the blood-vessels **in the abdominal** and pelvic cavities, resulting in painful and profuse menstruation, bearing **down,** back-ache, congestion **of the uterus and** ovaries, **a** partial suspension **or** an excessive **menstrual** flow, which in most cases will be followed by uterine ulceration.

Hemorrhoidal, ordinary piles and kindred troubles usually result from pressure, causing engorgement of the blood-vessels and consequent bursting of the delicate coats of the capillaries in the mucous linings.

And where gentlemen suffer from piles, hemorrhoids, and varicose veins, it is in all cases **the** result of pressure, **either** caused by constipation of the bowels, a rapid accumulation of fat **in** the abdomen, standing or walking **in** excess, frequent use of cathartics, all of **which** tend to obstruct the circulation in the main blood-vessels.

The absurdity of trusting implicitly to drug medicines to perform a cure in any of these cases must present itself to every thoughtful mind.

The first step to be taken is to restore the parts to their normal position, as we would do in replacing dislocated joints, when in most cases, if hygienic laws are observed, a cure will be established.

The treatment must at first be mechanical. When the walls of the abdomen have lost their tonicity, compresses must be worn, perfectly adjusted, so as to fit easily and snugly over the lower part of the body. These should always be used while walking or standing until the muscles have become contracted and strengthened. A daily deep hip bath, strongly impregnated with salt, should be taken, rolling and kneading the walls of the abdomen, but never rubbing downward. Instead lift and press the bowels gently upward, and form the habit of resting as much as possible on the face or stomach—Mussulman fashion—this being the only position (owing to the imperfect construction of chairs and beds) that gives the spinal column perfect rest, and at the same time removes the pressure from the large arteries and veins and sensitive nerves that lie against the spinal column and in the pelvic cavity.

Instead of sitting down to rest, as so many are in the habit of doing when fatigued, it were better to lie down, with the hips elevated six or eight inches higher than the shoulders, remaining in this position five or ten minutes, and assuming it several times during the day, and by degrees obtain control over the abdominal muscles, contracting them at first very gently, but persevering in the movement several times each day, while in a recumbent position.

No living room is completely furnished which does not contain one, or two if the family is large, commodious couches not too fine for constant wear, on which the various members of the family may recline while resting instead of sitting in ordinary easy chairs. Especially do we recommend these to the female members of the family. A few minutes rest stolen thus from time to time from the duties of the day will be more beneficial than as many hours in a sitting position, as there is a complete relaxation of all the muscles and a more perfect circulation.

An excellent apparatus, and one quite common in the health institutions, is a smooth, hard couch, elevated at one end perhaps twelve inches. On this the patient lies upon her face, with her head to the incline, and while in that position raises her body slightly up from the couch, holding her feet down, which are usually passed under a strap. An apparatus of this kind could easily be improvised in any house; an ordinary table, or a board with a quilt thrown over it, and inclined at a proper angle. While lying in this position full, deep breathing would be advisable.

The benefits derived from an elevating of the pelvic cavity in cases of misplacement of the uterus, especially prolapsus, has always been recognized by the medical practitioners. As long ago even as the time of Hippocrates, Euryphon suspended his female patients by the feet for several hours at a time.

Where the vaginal walls have lost their tonicity, tonic washes and the application of various astringent remedies may be resorted to. In this connection we can recommend Dr. Swan's Uterine Pastilles as a convenient and excellent tonic. Also as an assistant to Nature, a support in form of a good-sized bit of surgeon's sponge, which, after being carefully cleansed, may be saturated with glycerine and tannic acid, and introduced into the vagina, where it may remain for twenty-four hours, then be removed, by a cord previously tied through it, cleansed, again saturated and replaced, continuing this treatment in conjunction with that previously advised until the parts are able to sustain themselves.

We have never yet been converted to the use of the various pessaries so highly recommended by many practitioners. We consider that the cases would not exceed one in five thousand where their use would be justified. In extreme cases they come in **play as a** crutch would in case of a broken limb, to **be** used only until the parts unite and are healed; but for the common misplacements that are so prevalent, and due, as they are, in most cases to prolapsus of the abdominal viscera, we consider the pessary **an agent** of **harm**, rather than of use. It belongs to that class of helps which should only be resorted to when Nature no longer has any power to help herself—in **cases of weakness**, and not in cases of irritation. The inflamed and congested body **of the** uterus resting upon the hard substance often **becomes** more inflamed **from** its irritating presence, and abnormal growths are induced. The sponge pessary has not **the** objectionable points which can be urged against those in ordinary use. **It** is soft, and at the same time rather healing of the two, acting as an emollient to the inflamed *os uteri* and the vaginal walls.

A restoration of the parts depends largely upon the re-instatement of the muscles, and this must be accomplished through the judicious use of cool hip baths and muscular exertion, by voluntary contraction of the abdominal fibres together with those **of the** vaginal walls. These contractions may be impossible at first, but by perseverance can be performed with the greatest **ease,** bringing relief and strength to the parts. The recovery may be slow, but in most cases it will be sure. One must have patience, bearing in mind that it is easier to pull down than to build up the system, and that it may require years to restore the wastes of a day, and that we cannot draw largely where we have but little capital invested. If we are endowed with but a small stock of vitality, then we must learn to husband our resources if we would escape physical bankruptcy.

This existence is the primary department of the great life-school. We have just so much to learn here before entering into a higher grade, as it is doubtful about our being enabled through favoritism, money, or rings to obtain promotion there. If through ignorance of natural laws our whole lives have been a continual series of demolitions and reparations of the *covering* of the soul merely, then there has been but little time for the cultivation of that part which lives after the body has decayed. Did not some wiseacre say that an ounce of prevention was worth a pound of cure? Let this be the motto through life, so far as the physical structure is concerned. Study economy as much in the wear of your bodies as you do in the wear of your fine carpets and furniture.

We believe that the soul will grow more symmetrical and harmonious while inhabiting a symmetrical and healthy body; and if that religious sect, who hold the opinion that the spirit body is the outgrowth of the material one be correct, then we have a double incentive to protect and carefully develop all of the bodily powers; just as the mother would who desired a healthy and symmetrical offspring. We often see bodies which seem to possess no souls; but we have never seen souls walking about and attending to the affairs of life without bodies. Therefore, you who so strongly cling to this existence and desire to retain your hold on material things, take care of your bodies else your souls will seek more fitting garments in which to clothe themselves. Every time that you over-tax your forces, you lessen your chances for health and long life.

PHYSICAL DEVELOPMENT.

It is a very common occurrence to find delicate women with distortion of the breast bone, from tight lacing in early life, or from wearing ill-fitting and heavy garments; and there is no doubt but what many of the pelvic deformities arise from young girls wearing high-heeled shoes, together with an undue weight of clothing resting upon the hips before the bones have become sufficiently hardened to sustain the weight and balance the body in the unnatural position which high heels give.

The great suffering and peril attending child-bearing in these days result almost entirely from the lack of proper development of the female form; for not only is there frequently a deformity of the pelvis and breast bones, but through compression the ribs are curved in such a manner as to lessen by several inches the capacity of the body, a consideration of the utmost importance to the pregnant woman.

There is a perfect mania among our country-women for long and slim waists, regardless of the natural conformation of the individual. "Only make us sylph-like and long and slim," is the prayer of the long and the short, the lean and the fat, utterly insensible to the incongruity and lack of harmony of proportion, so that in most instances a fashionably dressed woman, unless she be really an artist and uses her artistic skill in dressing, is a mass of angles instead of curves, which if Nature had her way all women would present. There would be no sharp jutting hip bones and square shoulders.

Woman, as Nature intended her, is always perfectly adapted to fulfill her mission. She has beautifully sloping shoulders that she may with her arms more perfectly enfold as mother and nurse; she has large hips, long and well-developed trunk, that she may with greater ease bear and give birth to her children. The perfectly developed woman presents only curves; whereas the perfectly developed man shows only angles; the more angles the better; we want in him square shoulders, and narrow hips. Sloping shoulders and broad hips would be a deformity, an indication of an unbalanced character, and a tendency to this development in boys should be obviated by all kinds of athletic sports, use of dumb bells and gymnastic exercises, etc.

We do not want to be accused of putting forth the old stereotyped and hackneyed tirade upon dress, fashion, tight lacing, etc.; we only ask a candid survey of matters as they stand; we wish to take the middle ground—to be just. Let us hold up the glass to Nature and see how far we have wandered away from her straight and narrow paths; for although she is a kind and loving mother she is still an inexorable judge, and demands to the letter of the law that the pound of flesh be given where it is due.

Our truest judgments of men and things are founded upon comparison. Supposing we take as our models Venus de Milo, Venus de Medici, or Powers' Greek Slave, than which nothing can be more perfect, as our guides. With these forms before us we will start out on an investigating tour, and after a candid inspection let us compare notes, and take our word for it, we shall not be far apart in our deductions.

The law of Harmony is eternally the same, and eternally opposed to that of Discord. If Fashion runs her lines parallel with Harmony, then Fashion is all right, and just the one to follow; but if on the other hand they fall on the side of Discord, why then we must cut her or suffer the consequences. Tight shoes and high heels bring corns, bunions, aches, pains, and cold feet; so many protests against following Fashion on that side; long, heavily trimmed skirts and tightly laced bodices bring side-ache, back-ache, body and soul weariness,—still louder protests in favor of swinging round the circle and keeping on the side of Harmony.

It seems only reasonable that the pinched and unnaturally small waist with the sharply outlined shoulders and hips should suggest to the student of Nature what the crippled foot of the Chinese woman or the distorted skull of the Flat-head Indian would: simply a deformity, a deviation from the line of Harmony; and the deduction must be that as the cramped and deformed foot is unfitted for its mission, walking, and the flattened and depressed brain for normal

25

thinking and feeling, so must the angular and dispro-
portioned body fail in the performance of its highest
duty, namely, an uncomplaining and painless obedience
to the soul's behest.

The laws of Nature never deviate a hair's breadth,
therefore there is sorrow and physical suffering everywhere; not because God wills it, but because man is
ignorant and constantly transgressing the unalterable
decrees of Nature.

This suffering will continue until there is a clearer
understanding of the laws controlling the physical
organism, a knowledge of which is easily obtained.
When properly simplified a child may perfectly comprehend it. It is only the environments that make it
appear inaccessible, and any woman of ordinary common sense with a limited amount of study can readily
understand all that would be necessary for the care of
her own health and that of those under her charge;
and no woman has a right, in the highest sense of right,
to become wife and mother who has not this knowledge.
It should be taught to every child just as the rudiments
of the commonest education are, and the time will come
when it will be as much of a disgrace to say that we are
sick or feeble or nervous, as it will be to say that we
cannot read or write or think.

Physical incapacity **is**, either **directly or** indirectly, the result of ignorance, and ignorance **is** daily becoming **more odious** in the eyes of thinking people.

IMPROPER MARRIAGES.

The deepest hells of **human misery** to-day are the outgrowth **of** unsuitable and **unhappy marriages**, being not only **a source of** misery to **the contracting** parties, but to the **generations** that **shall** follow. **It is the** secret of infantile mortality, sterility, **of** predisposed insanity, natural nervous excitability. scrofula, **phthisis,** latent tendencies resulting from a lack **of temperamental** adaptation **on the** part **of** the parents; **the different** temperaments giving rise **to** various pathological conditions.

Take a man and woman of equally **refined** nervous constitutions, with **large brains, small necks and** chests, weak muscles, **similar color of** hair, eyes, complextion and conformation **of** head; **let them marry.** and **if** children come, **they** will be puny **and short-lived** in most cases, intensely excitable, disposed to brain and nervous disturbances, in **fact all** acute diseases. These **are the** sleepless children, **who** keep the house in a continual commotion. They **are never a** moment free **from suffering;** every **nerve is inflamed and on a** tension; **for** that which was something **more**

than ordinary nervous excitability in the parents has culminated in a highly inflammatory nervous state in the child. These children, if they live to become men and women, are the ones who are **always** complaining **of their** nerves, and constantly searching, but in vain, **for** some remedy for "nervousness." These are the "perfect bundles of nerves," **and are** forever kicking out of the traces and being denounced by their more fortunate fellows who happened to be born something near as they **ought to** have been. Poor souls! one might with **the** same propriety curse **them** for their thin bodies, sharp noses, keen eyes, and fine hair, as for their restlessness and irritability of temper.

In this class of temperaments the erratic geniuses are found, who fret and worry the life out of the more staid and quiet class of persons. **It** has been said that genius **is** only **one** remove from insanity; and the saying appears not **far from the** truth! In the above marriage there was a mingling of the two positive elements, giving rise in the offspring to the extreme development of the nervous temperament. The parents may have harmonized intellectually and affectionally; but physically, never! Such people, while really loving, yet irritate one another until life seems hardly worth living **for.** Marriages of this **class** are frequently fruitless. Such **was** the union **of** Napoleon and Josephine.

Standing in the attitude of brother and sister, or friends, their relations would have been the most amicable, beneficial, and lasting; but as man and wife they were unsuited to each other, as the sequel demonstrated. They made a *mistake*, just as thousands are doing to-day.

In opposition to the marriage of brains we have that of flesh in those of the lymphatic temperaments, the offspring of which present an excess of lymph, stupidity, indolence, and disease, with a predisposition to glandular disorders of all kinds. In the first example we had too much fire, too much concentration; in the second we have too little. While the nervous children need to be constantly held in check to prevent over-excitement of brain, those of the lymphatic constitution require to be as continually urged on to avoid imbecility. To eat and sleep, would constitute the lymphatic's dream of heaven. The parents of this class of offspring are no more harmonious in their relations with each other than are the over-wrought nervous temperaments. Their cold, inert, passionless, apathetic natures create mutual disgust and annoyance.

Then occasionally we have a marriage of two sanguine temperaments. In the children of such a union look out for red heads, and red-hot tempers, and all inflammatory diseases, scrofula, scrofula consumption, hip and

bone diseases generally. We have here a climax of the
combined heat of both parental temperaments. The
contracting parties do not as a rule live quite as we
imagine the angels do. On the contrary, they strike
fire like two flints whenever the contact is severe enough.
They can no more blend than two electric currents
could. They may regard each other's characters with
the deepest veneration, but as for living together harmoniously and peacefully they never will. It is simply
a physical impossibility.

Now and then we see a union between two unadul-
terated bilious temperaments, those with the black hair,
eyes, and dark skin. Then we see manifested an intensified sombreness in the children, except where Nature
overleaps herself and gives us Albinos. In this class we
find the stolid people who can have teeth extracted and
limbs amputated without flinching. Of this type are
many of the hardened criminals, those who have committed the most cold-blooded and premeditated murders,
to whom the ecstacy of joy, fear, or sorrow is unknown.
Men of this stamp walk to their death as they would to
their places of business, with no more apparent emotion.

The positive and negative always attract each other.
This is a fixed law by which harmony is preserved,
and were it obeyed in the selection of companions,
instead of mercenary and passional motives, there would

be more happy homes and fewer applications **for divorce.**

Attraction and **repulsion** govern *all* **friendships. Weakness coalesces with** strength, **not** weakness with weakness, **nor** strength **with** strength. Two positive natures, however well **balanced** and perfect in **themselves,** could **rarely** form lasting **friendly** relations; for all strongly marked characters **are more or less** angular, and these angularities fret **and annoy both parties.** How much more galling and soul-harassing **must** such a partnership **be when** bound **by the** bands **of** wedlock where every fiber **of the** body **and** every attribute **of the** soul repels the contact, not because **the** parties **are not** mutually true, and good, **and** noble **even, but** because they are irresistibly **thrown from** each other **as two** thunderbolts would **be. The strongest** chain **of** words **ever** forged by man **can** never change **this law** and bring **two uncongenial natures** in coalescence with each other, without **which no** marriage **is real.**

TEMPERAMENTS.

The four primary temperaments are distinguished by **the** shape of the head, size and proportion of the body, **color of** the hair, eyes, **and** complexion; more particularly by the conformation of the **head.**

In the sanguine is **manifested** firm, flexible muscles, broad chest, **large** heart, arteries and **veins, a rapid**

pulse, a full volume of blood, fresh color, the head
proportionally smaller than the body, and *flat at the
back*, well developed above the ears, but a sloping forehead. In this temperament we may find blue, gray,
and even brown eyes, fair, red or brown hair. This
represents the vital power and energy.

The bilious represents the sturdy side of humanity.
We find large bones, compact muscles, sinews of iron,
a brawny hardiness, the head is *well developed* at the
back and above the ears, but like the sanguine has the
receding forehead. The hair is coarse and abundant.
This represents the enterprise of the world, while the
sanguine symbolizes the steam.

In the encephalic, or nervous, the brain development
is in front of the ears. This represents the brain power,
the emotional and reflective capacity. Here we find
small bones, weak, flabby muscles, narrow chests,
small arteries and veins, a general lack of vitality
and force; and it is only when there has been a blending of this temperament with the two previously named
ones, that the front brain is pushed upward and forward
from the ears.

The lymphatic gives us a negative temperament and
opposed to the angular, which are displayed in the first
three. The head is round and smooth with an average
development in all directions; the bones and muscles

are weak, the adipose tissues predominate. This is the embodiment of weakness.

According to Prof. Powell, the sanguine and bilious are the original standard or vital temperaments, belonging to all people; the lymphatic and encephalic are the non-vital and the out-growth of civilization and cultivation; the lymphatic arising from conditions which wealth, luxury and idle living would induce; the encephalic from over mental activity, care, anxiety, sedentary habits and insufficient nutrition.

So far we have only described the extreme development of the four temperaments—Sanguine, Bilious, Lymphatic and Encephalic. By a proper cross in the parents we obtain a great variety, in all fourteen strongly marked constitutions, to say nothing of the infinite number of shades produced by the difference in the modes of life. Each of these constitutions vary according to the proportions of the standard temperaments in the individual. The sanguine may exist in the proportion of three to one, or the bilious may be in the ascendency, or the encephalic may predominate, or there may be an equal division of the three. The cross is rarely the same in any two, but it is an easy matter for the student to determine which is in excess and how many enter into the combination. This conclusion can be arrived at **by** a study of the *shape of the head*, texture of the hair and skin.

The best combinations are those where the four temperaments are united in equal proportions. Of such **was** Goethe, Napoleon, Agassiz, Macaulay, Thackeray, **and** among our marked men and women of to-day, M. Gambetta, Emerson, Robert Ingersoll, Rev. Henry Ward Beecher, George Eliot (Mrs. Lewes), Rosa Bonheur, Harriet Hosmer. Scores of others could be mentioned, but the above examples will be sufficient for **an** illustration. Nearly all men and women who have greatly signalized themselves **have** been of **that** complex organization. This crystallization **of** the **four** temperaments presents a wider mental range as well as a more perfect physical balance.

The sanguine gives vital force; the bilious, stability and enterprise; the encephalic, the emotive and intellective capacity; the lymphatic serves as an equalizer to **the** irritability of the bilious, undue force of the sanguine, **and** prevents over-excitation and consequent exhaustion of the encephalic.

The great importance of a proper cross in the temperaments of conjugal **partners** must be seen by even **this** brief survey of the subject. Sameness of constitution must always be avoided, not only for the sake of the progeny, but also to equalize the asperities and inanities in the companions. If this law was strictly obeyed for a couple of generations the world would be revolutionized. The current of habit is so powerful and yet so easily turned from its course.

Physical changes arising from some prevailing custom or fashion in one generation become fixed physiological facts in the next, as in the manifestation of the arched instep resulting from the progenitors wearing high heeled shoes, and as tight lacing on the part of the mothers abridges the breathing capacity of the coming generation. Galton says—"Each generation has enormous power over the natural gifts of those that follow; and it is a duty we owe to humanity to investigate the range of that power and to exercise it in a way, that, without being unwise toward ourselves, will be most advantageous to the future inhabitants of the earth."

CHAPTER VIII.

MATERNITY.

THERE is a mawkish sentimentality—a false modesty entertained upon the subject of maternity utterly at variance with common sense, and wholly unworthy of enlightened men and women. Not only is it unjust but manifestly vulgar; for to the pure all things are pure.

It seems incredible that men and women can find it in their souls to degrade by thought or act these Heaven-assigned functions, but men have trampled them in the filth of sensuality, until women come to regard them with disgust and loathing; not because this estimate is just, or womanly, but because men have been unjust, and hidden the God-principle under lust and degradation, and then cried, "Unclean!"

The low significance attached to gestation has a demoralizing influence, not only upon the mother and offspring, but upon all classes of society. It leads to many wrongs, to low, vulgar insinuations and speculations, it debars the pregnant woman of that freedom which is essential to the perfect development of her child; in fact, it embarrasses, distresses, and hedges her in on all sides. This undertone of mock modesty is undeniably low, and doubtless is the outgrowth of ignorance upon this important and interesting subject.

God has endowed man, and for a wise purpose, with those organisms which are intended to perform the highest and holiest of all missions, that of perpetuating the race—which must be a noble work since He made man in His own image, and pronounced him good— and since through His fixed and unalterable laws the propagation of the species has gone steadily on from the commencement of Time.

Every soul clothed in the flesh is a scintillation of the Almighty; a spark of the Divine Life; a thought emanation. What a cause for pride and exultation, as well as humility for the mother. Every woman who bears a child stands in the holy place where Mary of old stood, and likewise bears beneath her heart a portion of the Divine life; not mother alone to the fleshly body which dies, but to a soul that lives forever.

Our deepest sympathies are with those women who are compelled to live and die never having known the wholesome joys of maternity. They have failed in their highest life work; they have missed the most refining and exalting of all the processes of nature; for no woman can pass through that change without becoming stronger, purer, and nobler. Many **idle,** frivolous, and even vicious **women** have emerged from this ordeal baptized **with a** new life in the birth of the little soul. It **may be** the child **of** shame, poverty or crime, but it **is a** Messiah to her who has given **it** life, and though this child be ever so low in the scale of being it is still the child of God.

A belief is entertained by the Persians, that those women who have not borne children are never admitted into Heaven, but are compelled instead to stand at the outer entrance and hold open the gates while the sanctified and happy mothers, with **the troops of** joyous children, pass through into Heaven rejoicing.

"SHE WHO **ROCKS THE** CRADLE RULES **THE** WORLD,"

is an aphorism of deep significance, but she who purifies the hidden springs of her child's inner being, **as** she hour by hour gives it life, does even more. Through **the** refining influence **of** desired and wisely ordered maternity, the race will become improved **and elevated.** A more perfect generation will **do away with** the necessity **of** regeneration. Then in due time will appear the Coming Woman, and following closely, the Model Man, and then—the Millennium. Therefore, **we take** courage, although the population of the world is daily reinforced by half-made-up humanity, the children of chance and accident; for, where one child is the result of love, care, and forethought, thousands are accidental, and the outgrowth of passion merely, coming under the protest of both father and mother—and, so coming, fill the world with discord and bodily disease.

Unfortunately, legal marriages are not always real love marriages. Scores of men and women bound in wedlock are as much strangers to each other as though they had never met, the only point of union being the sexual act, and this often takes place under protest on the part of the wife, there being no love to prompt a desire. But this fact does not prevent her conceiving, if the condition of the uterus is favorable. She neither desires the act, nor the condition which the act imposes; she feels that her wishes have not been consulted, that she has been outraged, and, if conception takes place, she passes through the period of gestation with a smothered sense of wrong and indignation; momentarily shaping the destiny of her unborn babe, giving to it a life in which the springs are all inverted.

All well balanced and harmoniously organized men and women are the offsprings of a marriage of love. The children of love, though born of delicate and even diseased parents, have a better chance for a harmonious existence than those born of lust merely, though of the most robust parents. The spirit of love strengthens, upbuilds, and renews body and soul, while an indulgence in the grosser passions without love demoralizes the physical and spiritual nature of both mother and child.

Thus, through pre-natal influence, existence is rendered a burden to one-half of humanity. The cry is constantly going up, "I wish I were dead," or, "What is the use of living?" or, "What have I to live for?" The mere act of living, with many, although surrounded by every comfort, is a continual sorrow. Chateaubriand expresses the feeling of a large portion of mankind when he says of himself, "My life itself is one long weariness;" and again, "There are some intelligences that are half dead; mine was born so." Homicidal and suicidal tendencies are born with people, requiring only time and favorable circumstances to develop the germ of self-destruction and murder.

How this wonderful enfoldment takes place no one has ever been able to tell. We only know that the fact exists, and is doubtless in obedience to the great law of evolution. No one has ever been able to explain why the perfume of a heliotrope differs from that of the rose, or why one rose bush should bear white blossoms while another resembling it perfectly in outward form should bear red ones; or why one tree should be crowned with a wealth of sweet and luscious crimson apples, while the one standing close beside it, subject to the same general laws, should bear only sour white ones; or why two children born of the same parents, reared under the same happy influences, should show such totally opposite characters; one growing in grace and beauty while the other is as constantly developing vicious traits not manifested by either of the parents.

Such depravity can be accounted for only in one of three ways; first, on the ground of heredity—the cropping out of some remote ancestral proclivity which has not been sufficiently pruned and kept back in the parent stock, just as we sometimes lose our magnificent grafted roses because we become too ambitious for an abundant growth. When we do not use the knife enough, we are soon overrun with the coarse and vulgar growth of the original stock, giving us for our pains nothing but thorns, and inferior, inodorous blossoms. We have here a retrograde movement in nature, —a return to the original, and as every family has had, like empires, its rise and fall, there is no doubt but that there has been, along the ancestral line, rank briar bushes and crab-apple trees that will every now and then force out a vigorous shoot which will give much trouble before it can be pruned away, and at times is powerful enough to run the cultivated graft out entirely.

Second, that the mother's surroundings, mental and

27

physical conditions during pregnancy, would mould and influence, not only the entire future of the child, but to a degree, the coming generations.

The third proposition is that the offspring would be lastingly affected by the mental and physical conditions of the male at the time of coition: That the spermatozoa would be so impressed by the temporary conditions of the individual as to affect it in its enfoldment in the new being, and impress that being for good or ill during the whole life. The accumulation of facts elicited from a careful study of reproduction in man and the higher order of animals fully sustains this opinion. There can be no doubt but what any powerful mental or physical excitement of the male just before or at the time of the reproductive act, will, in nearly all instances, if conception results, give a decided bias to the character of the offspring; and in that way marked talents and idiosyncrasies may be accounted for when the condition of the mother has been normal and passive during the gestative period. The conditions surrounding the male at the time may have been abnormal and the result of accident, but the impression has been made as unerringly as that upon the photographer's plate. What are known as birthmarks are often given at the time of coition.

The mania for stimulants in some cases can only

be accounted for in this way: by the father being at
the time of the act under the influence of alcohol.
One instance comes to our recollection of a young man
who died of delirium tremens in his eighteenth year.
The father and mother were Scotch people, both
temperate and pious, but once, upon an occasion
of a reunion of his countrymen, he became con-
vivial, although not greatly so, and while under this
excitement cohabited with his wife, and the unfor-
untate boy was the result of this union. The man had
naturally no love of stimulants, and never indulged in
them even on great occasions until this one. The elder
children were models of excellence in every way. This
boy, being the last child, had everything in his favor so
far as maturity and development of character in the
parents were concerned. But he was the result of a
fatal accident. A momentary and thoughtless act on
the part of the progenitors caused a calamitous catastrophe to a blameless soul. One thus environed elicits
the pity and tears of angels, and the deepest sympathies and assistance of all men, instead of condemnation
and reproach. Crime is simply disease! And oh!
how broad your charities must grow when you learn
that *men are made*—that they do not make themselves
any more than the rose makes its color or the heliotrope
its perfume. Surrounding conditions evolve the latent

powers of the germ, but do not make it, and when you
see human beings stumbling and falling on all sides,
withhold your harsh judgment until you can know
something of the deep springs of their origin. Every
soul lives up to the law that controlled its germ-life.
If it was begotten in discord, it will develop discord;
if in harmony then the life will be guided by harmony.
In the truest and broadest sense, man is not to blame
for his peculiarities more than the tree is to blame
for bearing red apples instead of white; he may by
an earnest and persistent endeavor so develop his better
nature that in time the pre-natal blemishes will disappear entirely. He may accomplish this in half a life
time, while another soul less receptive to good influences and molded by the evil effects of generations of
bad breeding will require ages in the next existence to
unfold sufficiently to see the beauty of truth and purity
and the wisdom of God, but unfold it surely will. All
the advancement made by man in all the ages since
time began has been in obedience to that law. It is
only a matter of time. That which appears an eradi-
cable evil to us to-day may in the future prove only
undeveloped good.

Upon our table just now lies a luscious ripe peach
with its rosy cheek turned temptingly in our direction,
and as we take it up and part the thick, rich meat from

the pit, we go back in thought to the early history of this delicious edible, and we there learn that in its original and uncultivated state it was very small, hard, bitter as gall, and very poisonous, unsuited in every way to the use of man. What has wrought this change? Evolution. An unfoldment accelerated by the proper condition brought about through the influence of man. He did not make the peach, but he assisted it to unfold more rapidly than it otherwise would.

This law of evolvement governs all matter and spirit alike. Bad men and women are undeveloped men and women; souls imprisoned in gross and diseased bodies, furnished by gross and diseased parents, or else given at a time when the conditions of the parents were most unfavorable to conception. Such conditions, although only temporarily imposed, are just as fatal in their influence upon the offspring as though they were chronic states of the parents. For example, a woman while suffering from a temporary attack of lung difficulty may become pregnant, the affection disappearing in the mother some time during pregnancy, but the child dies in infancy of consumption, evidently inherited, although the disease has never manifested itself on either side of the family, the mother becoming perfectly robust after the birth of her child. Similar results are often witnessed when there has been a temporary physical or mental derangement on the part of the father at the time of coition; the child manifesting precisely the father's symptoms, the mother having been in unusual good health while pregnant. There is a perfect reflex of the mental and physical conditions of the parents at the period of conception and during gestation. What might be only a temporary brain affection in the father, induced by over-work or anxiety, proves a constitutional predisposition toward insanity in the child; or what was merely an accidental cold on the part of the mother developes in the constitution of her child, a deadly disease terminating in premature death, and in this manner physical peculiarities, diseases, mental biases and idiosyncrasies are developed, taking the offspring widely astray from the parental line. It is this influence on the germ-life that developes mental and physical prodigies, and at times geniuses.

AFTER CONCEPTION.

The modifying effects of the mother's condition and surroundings during pregnancy must not be underrated. She is continually transmitting to her child, little by little, all she feels, sees, or hears. For the time the mother and child are one. Whatever adversely affects her, in the same ratio unfavorably influences the foetus. If during the earlier period of gestation, while the fœtal heart is forming, the mother should be forced to undergo any very trying and exciting scenes which would produce great heart disturbance, there would be imminent danger of producing in the fœtus some organic trouble of the heart. The same would be true of the brain, and through similar disturbances physical blemishes, known as birth-marks, are produced, also abnormal sexual appetites; for every portion of the fœtal organization is more or less influenced by extreme activity of the corresponding portion of the maternal organism. The mother has therefore the power to mould and influence the life of her child for good or ill, her mental and physical condition, to a degree, counterbalancing the advantages or disadvantages existing at the time of conception. Nature is always endeavoring to preserve a harmony. Were this not so, the world would be filled with monstrosities. The more cultivated and well-balanced a woman is the less liability there is of her producing abnormal and monstrous developments.

During the first months of pregnancy, while the bones and muscles are being formed in the fœtus, the mother should take a reasonable amount of muscular exercise, that the child may possess a more perfect physical development. In the latter portion of gestation the brain is developing, becoming more firm and capable of being influenced by the mental conditions of the mother. There should be now only enough physical exercise to keep the blood in free circulation and the bowels regular. But the brain must have plenty of good, wholesome thinking to do. There should be a reasonable amount of solid reading indulged in, and a careful digesting of what is read. The reflections should be of a deep and earnest nature. To the thoughtful woman every blade of grass will suggest a sermon, every star will lead her deeper into the knowledge of the mystery of being. This knowledge makes the world appear sweet and beautiful, and humanity dear, causing the small, harrassing environments to melt away and disappear like a mirage. Subjects which were all Greek and Latin to-day, will be as clear as noon-tide light to-morrow.

For the mind is like a deep well: the more we draw from it, the purer its waters become.

The pregnant woman stands in a sacred temple into which the spirit of Envy, Jealousy, or Censure should never enter. The beauty of her child is not influenced so much by what hangs upon the walls and otherwise adorns her home, as by the beauty and grace of her spirit; although fine surroundings must in time refine the character of man, but not so rapidly and surely as would the mother's innate beauty of soul.

We have observed in the several departments of physical life that the important processes which are therein evolved require an expenditure of a certain amount of vitality, as in digestion, cerebration, ovulation, etc., and we have also seen that if the forces are greatly dissipated during the processes, they will, of necessity, be imperfectly performed. Now, gestation is a process of evolvement which necessitates an immense expenditure of vital energy; therefore the pregnant woman should be free from any great draught upon her vitality. Excessive physical exertion must in all cases be avoided, if she would not rob her offspring of its due share of vitality. What is true in the case of animals, is also true in the case of man. The intelligent stock-raiser will on no account allow his brood mares to be over-worked; and while they are with foal, treats them with utmost kindness and care, far exceeding that which scores of pregnant women receive.

There would soon be a wonderful improvement in the human family if man would bestow as much attention upon the improvement of his own species as he now does to the cultivation of his stock. Stock-raising has been brought to a wonderful perfection. The law governing the perfect propagation of animals is well understood, but men so wise in these matters seem never to have conceived the idea that the animal in man was just as susceptible of cultivation as the lower order of animals. The benefits accruing from a proper cross in temperament, a selection of the fittest, from a careful regard to the adaptability in age, and the season of reproducing their young, are quite as decided in their influence upon the human family as upon the animal creation.

In the human species we have a marriage between the angel and the animal, producing a perfect man. If the angel predominates, then he fails to perform his mission acceptably.

"Wings for angels, but feet for men."

Man's physical functions are as purely animal as those of his lower kin, and the more implicitly he obeys the natural laws governing the animal functions the more robust and symmetrical he will become; he should not carry his animality into the realms of his higher nature any more than he should drag the angel side down into the animal domains. Each department has its own work, and one is just as sacred as the other. We have observed one significant fact in connection with the propagation of the lower animals,—a course pursued by them which it would be well for man to emulate, to keep him at least on a *level* with animals, and that is, that the female is never disturbed by the male during the period of pregnancy. Copulation is never permitted on one side, or demanded on the other. Furthermore the male makes no overtures to the female unless she courts them, which she does not, except when in a condition to conceive; and the act during gestation becomes no less a wrong and a physical outrage in the case of man than in the case of animals. Man has been endowed with reason to guide him and keep him above the brute. The angel nature was placed uppermost to hold in check the animal passions. The greatest wrong from which Humanity suffers to-day, comes through an abuse of the sexual nature.

Pregnant women, as a rule, are averse to the sexual union during the period of gestation, and if a desire should be manifested at the time, it may be regarded as the result of some abnormal condition; perhaps from ulceration of the womb, leucorrhea, granulation of the vagina, etc., and the case should receive medical treatment of a soothing nature instead of a passional indulgence which would tend to increase the irritation. The detrimental effect upon the child is manifold to both soul and body, and never beneficial, as many claim it to be.

Dr. Black says: "Coition during pregnancy is one of the ways in which the predisposition is laid for that terrible disease in children, epilepsy. The unnatural excitement of the nervous system in the mother by such a cause, cannot operate otherwise than by inflicting injury upon the tender **germ** in the **womb.** This germ, it must be remembered, derives every quality **it** possesses from the parents, **as well** as every **particle of matter** of which it is composed. **The old** notion **of** anything like spontaneity in the development **of** the qualities **of a** new being is at variance with all **the** latest facts and **induc-** tions concerning reproduction. And **so is** that **of a** creative fiat. The smallest organic cell, as well as **the** most complicated **organism, in form** and quality, **is** wholly dependent upon the **laws** of derivation. These laws **are** competent **to** explain, however subtle **the** ultimate **process may be, the great** diversities **of** human organization **and** character. Impressions from without, the emotions, conduct, and **play of** the organic processes within, **are** never alike **from day** to day, or from hour to hour; and it is from **the** aggregate of these in the parents, but especially **of** those **in** the mother, imme- diately before and after conception, **that the** quality of **the** offspring is determined. Suppose, **then, that** there **is** every now **and** then **an** unnatural, excited, and exhaustive state **of** the **nervous system** produced **in** the mother by excessive cohabitation; **is** it any wonder that the child's nervous system, which derives **its** qualities from those **of** its parents, should **take its** peculiar stamp from that of the parent in whom it lives, moves, and has its being? In the adult, epilepsy is frequently developed by excessive venery; and the child born with such a predisposition will be exceedingly liable to the disease during its early years, when the nervous system is notoriously prone to deranged action from very slight disturbing causes.

"The infringement of this law regulating intercourse during pregnancy also reacts injuriously upon the mental capacity of the child, tending to give it a stupid, animalized look; and, there is also good reason to believe, aids in developing the idiotic condition."

A serious injury is also inflicted upon the pregnant woman by inducing irritation of the uterine nerves, increasing the peril and suffering of child-birth, as well as retarding convalescence, she requiring all her strength to carry her through the ordeal which she is to pass.

An excess of coitus brutalizes man and degrades woman; and in most cases is the death of love in the marriage relation. The demoralizing effect shows itself in those male animals kept for the purpose of propagation. A degree of viciousness almost amounting to madness is displayed by stallions and bulls that have been used in excess for a length of time. A false idea prevails regarding these functions. The sexual, like all other appetites, increases from an over-indulgence. From over use and consequent excitement, the energies become focalized in the direction of the sexual functions.

Marriage in most cases is simply a license to unbridled indulgence, which brings physical and mental bankruptcy. If men could be convinced of the important physiological fact, that an excessive loss of semen is just as destructive to physical, mental, and spiritual upbuilding, as a daily drainage from the arteries would be, a world of suffering would be prevented. The seminal fluid is composed of the best arterial blood and a large supply of nerve force which would go to the reinforcement of brain and muscle.

Dr. Gardner in his excellent work on "Conjugal Sins," says, "The sperm is the purest extract of the blood.   *   *   *   Nature, in creating it, has intended it not only to communicate life, but also to nourish the individual life. In fact the re-absorption of the fecundating liquid impresses upon the entire economy an entirely new energy, and a virility which contributes to the prolongation of life."

Parise, the noted physiologist and microscopist, thus writes: "Nothing costs the economy so much as the production of semen and its forced ejaculation. It has been calculated that an ounce of semen was equal to about forty ounces of blood. Semen is the essence of the whole individual. It is the balm of life. That which gives life is intended for its preservation.''

Newton ascribed his power of mental concentration to the fact that he lived a life of total abstinence from sexual indulgence. What in most men would be lost in the seminal discharge, he called to, and used up in his brain; and there are scores of well authenticated cases where men with strongly marked sexual natures have lived celibate lives, and have steadily improved in mental and physical stamina; not from a loss of power, but from holding the passions in abeyance to the higher nature. Vigorous and healthy brain work will reduce undue excitement of the genital organs, even when that excitement is largely due to a diseased condition of the organs or an irritation of that part of the brain governing them.

Sexual excitement is largely the result of the emotions, and is as readily controlled as aroused. Physical contact suggests the thought; thought gives impulse to the bodily organs, and thus excitement is produced. If married people would occupy separate beds there would be no more difficulty in controlling the excesses after, than before marriage.

Scores of weak nervous women are called upon to bear a child every two years, and yet during the period of gestation grant their husbands a weekly, if not a daily sexual indulgence. Such mothers go on without a protest, giving birth to weak, puny and diseased children, and so the world groans under its burden of disease, simply because man continues to gratify his passions regardless of the suffering that must accrue to others. Delicate women after bearing children should have, at least, a year's rest from a draught of that nature.

When men grow wiser and less selfish, and become as willing to expend as much time and money upon the cultivation of their higher faculties as they now do to feed and pander to their lower passions, there will be fewer widowers to be consoled, delicate broken-down wives to be medically treated, feeble and sickly children to suffer and die. Then too, houses of prostitution will cease to be licensed as institutions necessary for the protection of virtue. We see no way out of this labyrinth than by properly educating the youth; boys, to control and govern their lower nature, to overcome selfishness which would seek gratification at the expense of another's happiness, and to esteem their bodies as sacred as their souls; and girls to fully appreciate a woman's prerogative, *the full control of her person,* a control which no marriage contract, however binding, can take from her; and as shocking as it may appear, we shall say that this is woman's work. Her mission is to boys, as well as girls, and no woman should allow false modesty to stand in the way of her performing this duty. It matters not whether it is sister, mother, friend or guardian, it is the *woman* who must impart these truths to the young, that the coming generations may escape the mistakes and suffering of this.

CARE OF PREGNANT WOMEN AND THEIR TREATMENT AFTER CHILD-BIRTH.

After conception, the clothing, if not already so arranged, should be supported from the shoulders where the pressure can do no other harm than give a temporary inconvenience at most; then all of the garments should be loosened about the waist so that the uterus, as it enlarges and rises, may find no obstruction in its way. This rising of the organ takes place about the fourth month, unless the clothing be too heavy, or worn too snugly, in which case the uterus does not rise as it should, consequently great suffering is experienced from pressure upon the bladder, rectum, kidneys, blood vessels, arteries and nerves. This condition gives rise to bladder and kidney difficulty, constipation of the bowels, varicose veins, cramps in the lower limbs, hysteria, and at times a mild type of insanity; but where the abdominal walls are flexible and expand, and the fœtus rises, none of these inconveniences are experienced, the health usually improving from the first,

29

extremely delicate women often becoming strong and
hearty, and when the proper treatment has been given
during convalescence the benefit remains permanent,
for if all things work normally, child-bearing is most
wholesome in its ministry, and we rarely find women
healthy in the genital organs who have for any length
of time prevented conception. It is an outrage which
Nature protests against, and which gives rise to hardening and enlargement of the uterus and a general
disturbance of the menstrual functions.

When prolapsus of the bowels and uterus have existed
for a length of time before conception has taken place,
it would be advisable to wear perfectly fitting abdominal bandages wide enough to support the whole lower
portion of the trunk. These supports will be of great
comfort and benefit, as they will assist the impregnated
organ to rise, and also prevent it from settling down
while standing or walking. We advise this aid when
the walls have lost their tone and are incapable of
sustaining the weight of the fœtus, just as we use props
and frames to support a tree when it is overloaded
with fruit. By this support the woman is enabled
to take the necessary exercise to ensure health to
herself and her offspring, for nothing can be more
detrimental than lack of exercise. An excess of
bodily exertion is even preferable to a freedom

from activity. This bandage, arranged with broad straps passing over the shoulders, so adjusted as to support the weight of the abdomen, will prevent the dragging sensation so distressing toward the latter portion of the period. The higher the burden is carried the less the pressure on the veins and arteries, and the greater the freedom in the use of the limbs and body generally. A bandage should be worn for several weeks after convalescence, until the uterus has gained its usual size and the ligaments, peritoneal and muscular walls their normal tone. If these precautions are observed, and bands of every sort discarded, there would be fewer cases of uterine misplacement after child-birth.

Regarding the diet during gestation, much has been written and still more said, but no absolute guide for general use has ever been satisfactorily presented. The regimen advised for this class of patients seems to have been the result of opinions merely, and not based upon scientific facts. One advises an exclusive use of farinaceous food, another claims that a too free use of breadstuffs tends to harden the cranial bones in the foetus, and prescribes little bread and a large amount of fruit, so that the bones shall not be well formed and the labor thereby rendered easier. It is quite as difficult to arrange a bill of fare for pregnant women, as a class, as it would be to do so for the masses. What would best suit and nourish the mother would most perfectly develop the fœtus. The quantity more than the quality is to be considered. So long as it is of the best and wholesomely prepared, and meets the demands of the mother's organism, the variety **is a** secondary consideration; **for** observing **a wise** law the fœtus attracts to itself what it most needs, and if there is a dearth of certain principle in the blood, **the** mother will be more likely **to** suffer **than the** child. We have found this true in numbers of instances where experiments in dieting **have been** tried and breadstuffs largely dispensed with, **but the** bones **of** the fœtus were found **to be as firm as in** the previous cases where this precaution **had not** been observed.

Over-eating is a wrong which must be most carefully guarded against, especially during the earlier stages of gestation, and in fact, this caution is necesary during the entire period, unless the *enciente* woman should be compelled to perform a great amount **of** physical or mental labor. The temptation **to over-eat** will in **most** instances be very great, resulting **in** aggravated suffering from plethora, over-development of the fœtus, accumulation of flesh about the genitals, all contributing to prolonged and painful parturition and greater danger of fever and inflammation during the child-bed period. **The** food must be substantial and of such a nature as to act gently on the bowels. Upon no account allow constipation to exist for any length of time. Where there is a tendency of the system to secrete too great an amount of fluid, less liquid should be used, but a varied diet in all cases would seem best, and as a rule fancies, when the articles desired are not in themselves unwholesome, should be indulged, as the craving of the appetite is often a demand of the system for some substance or subtle spirit needed by the building forces. The only arbitrary rule that should be enforced is that of temperance in all things. Monstrosities in appetite can always be controlled by a matured mind and a strong will; in fact, rarely appears in a well developed and harmonious woman.

In order to secure a healthy and abundant secretion of milk, the mammary glands should be bathed with cold water, and very gently manipulated and pressed after each bath. This treatment must be persevered in daily from the commencement to the close of the period of gestation. This treatment will also act as a preventive of sore nipples. Vaseline should also be freely used to anoint the breast after bathing.

During the last three months of pregnancy a warm hip bath should be taken every alternate night upon retiring. Have the bath so arranged that it will not be lower than an easy chair. On no account let the position be cramped. **From** the third month anoint the abdominal **walls** and external genitals and vagina daily with vaseline, that they may become flexible and easily expanded during **the** gestative **and** parturient **period.** The hip bath **will** facilitate **labor** and hasten convalescence, and should **be** frequently administered after the birth of the child, observing **great** caution that the patient does not exert herself too much in **the effort of taking them.** The **baths should be warm** enough to prevent a sense of chilliness.

DUTIES OF THE NURSE.

It is the **nurse's duty to gently overcome any** dislike **or fear on the part of** her **patient to such a bath.** Let the first bath be quite warm, avoiding any dampness of the clothing **when the patient** retires **to** her bed. **A hot** brick or a soap-stone is **as** much **an** indispensable **in a** house where there is sickness, **as a** good doctor or nurse, and particularly in the cases of confinement.

The diet of the convalescing **patient should be the same as that** used during **pregnancy, only here again we must** guard against over-eating.

Change **and purify the** air in the sick **room,** renew the body **and bed** linen often, and cleanse the patient **by** frequent **local** bathing. Varying the position from lying to sitting assists the uterus **in** its effort to contract and disgorge the over loaded blood-vessels. Puerperal fever is more frequently the result of improper care and lack of thorough cleansing than from any other cause. The horizontal position in which lying-in women are usually kept for so long a time prevents a perfect drainage. The unreasonable and pernicious practice of keeping fresh air from rooms of this class of patients, cannot be too strongly condemned.

Fancies and superstitions should be expurgated from this branch of therapeutics. A nurse controlled by the prevailing whims and notions is unfitted for this work. She should be as carefully and thoroughly educated for her post as the physican for his. In the cases of delicate, nervous and imaginative women, the recovery is often retarded through fear and anxiety created by the superstition of the nurse, or the women who are present at the birth. If the parturient woman lifts her arms above her head, she is told that it is a sign that she is not going to recover. If she sneezes on Friday morning before breakfast, she will most surely lose her baby within two weeks; and if she dreams of catching fish it is a sure omen that the whole family are speedily to be stricken down with small-pox, and so on to the end of the catalogue of absurdities. Not only should a woman have calm and pleasant surroundings during child-birth, but also throughout her convalescence ; and we may say, the entire period of nursing, as she still greatly influences her babe through her milk. She should be free from all mental excitement of an unpleasant character, also any unusual strain upon the physical **system, in fact** all excesses must be avoided until nature restores the draught made upon the organism, through gestation and parturition.

CARE OF NEW-BORN INFANTS.

It must be understood that the infant now has a new element to battle with. **It** is compelled to breathe for itself. The rudimentary air-cells have never been expanded until now; the blood **has not** entered the **lungs for** purification, the nutrient blood having been supplied from the mother's circulation, the communication being through the arteries in the uterine walls forming a union with those in the umbilical cord and placenta, the latter attaching itself to the interior walls **of** the uterus very soon after conception takes place, thus establishing a free circulation between **the mother** and child, and by this mean the latter obtains **its nour-**ishment; just as **a** young tree **by** penetrating the **earth** with **its** roots and rootlets draws its life therefrom.

There is a muscular partition dividing the heart into **two** parts known as the right and left sides. After birth the right **side** receives the impure blood and impels **it** into the lungs, where it is freed from carbonic acid gas, while the right side receives from the lungs the pure or arterial blood and forcibly drives it into the arteries; but in the fœtal heart there is a small valve in the mus. cular septum called the *foramen ovale*, through which the purified blood passes from the right into the left side and then into the general circulation, not entering the lungs at all.

When the child is expelled from the uterus, there is instantly a respiratory impulse and a quantity of air is forced into the air-cells; the new being has established a communication with the outside world. Only a limited number of air cells, however, are filled in the first effort, and those not fully inflated.

The more lustily the infant cries, the more perfectly will the lungs be expanded. Each air cell is like a bladder which can be inflated to, comparatively, a great size. If the chest of a newly born infant were measured at the moment of birth, and then again twenty-four hours afterward, there would be found a difference of from two and a half to three inches expansion caused by filling the lungs with air.

It must therefore be seen that the practice of bandaging infants is an injurious one at all times. Many of the bones are scarcely more than formed, not hardened, and consequently afford no protection or resistance to the pressure. The breast-bone and ribs are often permanently distorted by this barbarous mode of dressing. Bandaging is no more necessary in the cases of infants than in that of young animals. There is less danger of weakness or breach, and the navel dries and sloughs off much sooner when bandages are loosely arranged. In our experience, the results of the following treatment of infants have always been highly satisfactory :

As soon as the cord is severed and ligatured near the body of the child, let the nurse be ready with a warm, soft, flannel blanket, and receive the babe, wrapping it so as to prevent the air reaching the body. It matters not what the season may be, if in mid-summer, there should be artificial heat in the room where it is first dressed. Let the nurse now carefully dry the body with a soft, old linen cloth, and then with her warm hand anoint the entire surface with vaseline, passing her hand underneath the blanket, and not allowing the air to reach the body. Where vaseline cannot be obtained, the oil from fowls — chickens or turkeys — would be preferable to the vegetable oils, as they are more penetrating and softening, and less gluey in their nature. Only a small amount need be used.

After this oil bath, the navel must be attended to; a square piece of fine old linen, with a circular opening made in the centre, must be oiled and put on, passing the cord through the opening; then pass loosely around the body, and secure in its place, a soft linen or fine flannel band, to prevent the cord from being displaced in changing the position of the infant, but it must be left so loose that the ribs will have free play as the lungs become expanded. Then a soft diaper and a simple little slip of fine flannel or cambric completes the baby's first toilette.

As soon as the mother is sufficiently rested, the child should be placed at the breast, receiving therefrom its first food. Nothing crude should enter the stomach of a new-born babe. There is stored by Nature, in the breast of the mother, that which cleanses the entire alimentary canal of the infant, and nothing should ever be substituted for it. The prevailing idea that new-born infants cry because they are hungry, has given rise to the pernicious habit of stuffing and dosing, which, in nearly all cases, lays the foundation for colicky habits. All substances except the mother's milk are too crude for the delicate stomach, and induce indigestion, and consequently, colic. The practice which many nurses have of dosing infants with sweetened water, gin and water, milk and water, as soon as born, is a most injurious one, and should never be allowed.

It was the opinion of old Dr. Mott that twenty-five per cent of all the children born died from the too sudden cooling off of the body immediately after birth, and where death did not supervene, organic difficulty of the heart, lungs or kidneys was frequently induced by the exposure of the tender, heated, and moist surface of the delicate body in the process of bathing and dressing. The temperature of the fœtus before birth is nearly that of blood-heat, while the sick room is not above fifty degrees, and many times it is necessary to have a still lower temperature, particularly where the labor has been very severe. The bathing of a limb at a time and allowing the cold air to strike the wet skin, tends to close the capillaries and drive the blood to the vital parts, the tender organization having little recuperative power to resist the effect of such exposure.

Scarce one babe out of fifty, bathed and dressed in the ordinary way, but receives a shock to the entire system which lays the foundation for a crying, troublesome babyhood. The long, inconvenient, and tight clothing tortures the child into fretting and crying from sheer misery, which is construed into some kind of sickness needing medication; then commences the dosing, and, as a sequence, sick and crying babies. There is no good reason why they should cry and fret continually, any more than that young animals should.

They are simply tortured into it. There was never yet a "cross baby." The world is full of sick, but never cross, ones.

Twenty-four hours after birth is quite soon enough to bathe an infant, and then the bath should be air immersion, and the temperature higher than that of the body, which should be kept covered with the water during the entire bath. Then have plenty of hot flannels to envelop the little body while in the act of drying it, on no account permitting the air to reach the moist flesh. The soap used should be the old, pure, white castile, and after every full bath anoint the skin with vaseline or oil as at first.

Always bear in mind two important points in the care of these little strangers, namely, that they will bear a great amount of heat, but at the same time are quite as dependent upon a good supply of pure, fresh air as the mother. They must now fight their own battles for life, and they do this against great odds while stowed away under a mountain of bed clothes, compelled to inhale the effete exhalations from the mother's body. It is truly wonderful that the tiny specks of humanity do battle so successfully; it is a marvel that they come through at all, so little is done to make them sound before they are born, then after they arrive here, they are compelled to combat the combined efforts of nurse and family relations to kill them with kindness.

We are never called upon to inspect a basket of baby fixings that we don't feel tempted to pitch the whole thing into the fire. If mothers would spend one-half as much time in reading, thinking, and cultivating their minds while carrying their children, as they do in the stitching, tucking, and embroidering these abominations to torture the little bodies after they do come, there would, as a rule, be presented a product more worthy of the time and strength consumed, and suffering endured during gestation and parturition.

"Love's love lost," seems a most fitting motto to place on the receptacle for these marvels of lace, muslin, linen, and costly embroidery, when we take into consideration the puny, half-made-up, diseased, and short-lived children who fall heir to these dainty things. The richest endowment to an earthling is a sound body and a harmonious soul. It is the *right* of every soul to be born thus, and if it is not so endowed, it has been robbed of its birthright. Never mind the rags that shall cover the child. Think only of the babe, and let it be a *child*, in the truest sense, and not an *abortion*. The world is full of such, who groan and suffer all through life under the weary load laid upon them by their progenitors.

We have been howling long and loud over heavy clothing and trailing dresses on women, but little or nothing has been said or done to improve the dressing of infants. The only step forward has been in the direction of increased ornamentation, which in itself is no doubt beautiful, but not needful for the requirements of health and the development of the child. There should be the utmost freedom to the limbs. The clothing should be worn perfectly loose, and yet so arranged on the shoulders, that nothing would pull down as the infant is moved about in being cared for. Keep shoes from the feet as long as possible, and when they are put on, never fasten them tightly about the ankle. The hand-made knitted shoes are far the best, as also are knitted shirts and belly-bands, because they are elastic and at the same time warm.

Another important consideration in the care of infants is to leave them to themselves as much as possible, that is, not to hold or handle them too much. Change the position frequently but avoid the habit of holding them constantly. A person of great nervous irritability should never have the care of young children, not even mothers, if they be delicate and of excitable temperament, because those things are communicated from mother to the child as it would be between adults. Old or feeble people should never sleep with or **fill the** position of nurse to young children, **as** it **proves most destructive** to the well-being of **the** little **ones**. The injurious effects may not manifest themselves outwardly for **a** time, but there will **be a slow** undermining of the general health, the elder absorbing from the younger, slowly sapping away the life.

Great care should be observed in **the** diapering of female infants especially, **to** prevent any distortion of the pelvis bones, as they are not hardened, particularly **the pubic bones.** Napkins should be discontinued as soon as possible, as the mass of clothing between the limbs has a tendency **to throw** them out of **a correct line,** destroying the **balance of** the body, producing the extreme of "toeing out," bow legs, halting gait, etc.

Infants can early be taught habits of cleanliness, **and** when they are fed at regular intervals, the passages will be correspondingly regular so that there **will be no** more necessity of keeping diapers on **babies than** on older children. It is only those who are too frequently fed **or** nursed that give so much care and trouble by their irregular bodily habits. We have reference **to those of** from four **to six** months old who would not require feeding oftener than from three to four times a day. A new-born babe would require to be fed oftener, perhaps as often as five or six times, as the stomach is undeveloped and would only contain a small amount of food at a time, and as the circulation is very rapid the digestion and absorption would be more quickly performed than in an older child. The crying babies as a rule are those that are always being "stuffed," the breast or bottle being kept almost continually in the mouth, because it is supposed that they are suffering from hunger when they are really crying from the misery of being over-fed, which induces indigestion, flatulence, and colic.

It is claimed by many that vomiting is a sign of health in an infant. It is no more a symptom of health in a child than in a full-grown man. It is simply an effort of nature to free the over-loaded stomach from its embarrassment. We should scarcely pronounce a man in health who would go from his table vomiting about the streets. He would get little sympathy we fancy. In the gluttonous days of some of the Roman emperors, there was attached to the banqueting hall, a room known as the vomitory, into which the banqueters would retire and produce the vomitive act, and then return and again resume the feast. Now we all pronounce this *beastly* in the extreme, and yet one-half of our babies are going through with this same system of cramming and ejecting, keeping them, as well as every-

31

one connected with them, in constant misery. Babies
should be no more trouble than young animals; all
they require is warmth, quiet, and a suitable amount
of wholesome food. Allow them to vegetate for at
least two months, then when the brain begins to
solidify sufficiently to receive and retain impressions,
give a little more attention, but do not overdo the
matter even then.

HOW TO DETERMINE THE NUMBER OF OFFSPRING.

However much may be said to the contrary, there
can be but little doubt in the mind of the earnest
student of Nature that the sexual act was designed for
reproduction alone, and where there is an indiscriminate
indulgence, with a steady determination to thwart that
object, it must, like all other transgressions, receive a
punishment due the crime.

During coition there is an influx of blood to the
uterus and surrounding organs, evidently for the purpose of assisting in conception if it takes place, which
if it does not, then the surplus blood and nerve force
attracted to the parts, and not thus consumed, must
ultimately lead to local derangements.

The detestable and pernicious practices so commonly
indulged in by married people is no less injurious to the
wife than to the husband. The more general mode of

prevention is what is known as the withdrawal system, a species of Onanism, both destructive to health and disgusting to a refined mind. In men of a nervous temperament, if long continued, it brings irritability of temper, depression of spirits, consumption, epilepsy, vertigo, and so called "softening of the brain," and drives hundreds to the verge of insanity. It is a substitution of artificial for natural conditions, an outrage for which Nature beats the offender with many stripes. Acton says:

"The excited nervous system, if it does not receive that shock which we have seen attends ejaculation, suffers a longer and more severe strain, lasting often days or nights, and one that is repeated over and over again. In fact, the non-occurrence of emission after sexual excitement, permits, for a time, the repetition of the excitement; but ultimately a collapse takes place from which it is very difficult to rally a patient. These practices, unnatural in the highest degree, cannot be carried on with impunity. Nature is sure, sooner or later, to inflict a severe retaliation."

In the woman, it induces engorgement, inflammation, and permanent hardening of the uterus, together with a full catalogue of the lesser evils. To quote from Dr. Francis Devay, we have the following:

"However, it is not difficult to conceive the degree of

perturbation that a like practice should exert upon the genital system **of woman by** provoking desires **which** are not gra**t**ified ; **a profound** stimulation is felt through the **entire** apparatus; **the** uterus, fallopian tubes, and **ovaries** enter into **a state of orgasm,** a storm which is **not** appeased by **the** natural crisis; **a** nervous super-excitation persists. There **occurs, then,** what would take place if, presenting food **to a** famished man, one should snatch **it from his** mouth after **having** thus **violently** excited his appetite. The sensibilities **of the womb and the** entire reproductive system are teased **for no purpose. It is to this cause,** too often repeated, that we should attribute the multiple **neuroses,** those strange affections **which** originate **in** the **genital** system of **woman.** Our conviction respecting them **is based** upon **a great number of** observations. Furthermore, the **normal** relations **existing** between the couple undergo unfortunate changes. This affection, founded upon reciprocal esteem, is little by little effaced by the repeti-**tion of an** act which pollutes **the** marriage **bed;** from thence **proceed** certain **hard** feelings, certain deep **impressions** which, gradually growing, eventuate in the scandalous ruptures of which the community rarely know the **real motive.''**

Mayer, **who** is perhaps one **of** the **best** authorities, **says:**

"If the good harmony of families and the reciprocal relations are seriously menaced by the invasion of those detestable practices, the health of women, as we have already intimated, is fearfully injured. A great number of neuralgias appear to us to have no other cause. Many women that we have interrogated on this matter have fortified this opinion. But that which to us has passed to the condition of incontestable proof, is the prevalence of uterine troubles of enervation among the married, hysterical symptoms which are met with in the conjugal relations as often as among young virgins, arising from the vicious habits of the husbands in their conjugal intercourse. * * * * * Still more, there is a graver affection, which is daily increasing, and which, if nothing arrests its invasion, will soon have attained the proportions of a scourge; we speak of the degeneration of the womb. We do not hesitate to place in the foremost rank, among the causes of this redoubtable disease, the refinements of civilization, and especially the artifices introduced in our day in the generic act. When there is no procreation, although the procreative faculties are excited, we see these pseudo-morphoses arise. Thus it is noticed that polypi and schirrus of the womb are common among prostitutes. And it is easy to account for the manner of action of this pathogenetic cause, if **we** consider how probable it is that the ejaculation and contact **of** the sperm with the uterine neck, constitutes **for** the woman, the **crisis of** the genital function, **by** appeasing the venereal orgasm and calming the voluptuous emotions, under the action of which **the** entire economy is convulsed.''

The various mechanical appliances employed, as well as the cold douches used to prevent conception, are also a prolific cause of uterine derangement.

Dr. Gardener says:

"**It is** undeniable **that** all methods employed **to** prevent pregnancy are **physically** injurious.'' In speaking of cold douches as a prevention, he remarks: "When in **the** general state of nervous **and** physical excitement attendant upon coitus, when the organs principally engaged in this act are congested and turgid with blood, do you think you can with impunity throw **a** flood of cold **or** even lukewarm water **far** into the **vitals in a** continual stream? Often, too, women add strong medicinal agents, intended to destroy by dissolution the spermatic germs, ere they have time to fulfill their **natural** destiny. These powerful astringents suddenly corrugate and **close** the glandular structure of the parts, and **this** is followed, necessarily, by a corresponding reaction, and the final result is debility and exhaustion, signalized by leucorrhea, prolapsus, and other diseases.''

In order to remedy these evils, men must learn self-control. Then so much plotting and counter-plotting to circumvent Nature will be unnecessary. In order to prevent a too rapid multiplication of the species, a wise provision has been made in the law governing reproduction; this law controlling all forms of animal life. It is a well established fact in the science of procreation that conception can only take place after a certain preparation has been undergone by the genital organs.

There is a portion of every month during which a woman would not be in a condition to conceive. The law of periodicity controls these functions. There is a time for the evolvement of the ovum, and a time for its expulsion, just as there is a time for a tree to put forth leaves and blossoms, after which comes the green fruit, then the ripe, then its expulsion from the tree. A chain of unfoldments and decadences. Nature makes her pauses, which are as marked and important as her periods of action. It would not be consistent with the law of motion and rest which governs the world of mind and matter that the uterus should be at all times ready for conception. We know that it is not. The time intervening between the expulsion of one ovum

and the ripening of another differs in different individuals, but the period is uniform in the individual,
occurring, as a rule, with great regularity. In some
instances the ovum may be expelled in seven days, in
others, eight, ten, twelve, and in rare cases during the
menstrual flow. When the muscles of the fallopian
tubes are not strong, which condition exists in the
majority of our delicate women, the time intervening
between the commencement of the menses and the
expulsion of the ovum would not be much short of
from fourteen to sixteen days, and not unfrequently is
retained until the eighteenth day; but every intelligent
woman can, by studying her bodily habits carefully, in
time tell just as readily when the ovum is expelled as
she can when the menses arrive.

During the passage of the ovum through the fallopian tubes there is generally more or less uneasiness,
and at times, extreme pain, from the cramping of the
muscles in the walls of the tubes. This is particularly
true in the cases of young girls who have been exposed
to cold during the menstrual period, but in all women
its passage is more or less marked. The fallopian
tubes, by their worm-like movement carry the germ to
the uterine cavity, where it remains for a time, the
period varying in different persons, at the end of which,
with a slight expulsive effort, it expels the deciduous

lining and ovum together, something after the manner of child-birth, leaving the uterus barren until another period of ovulation. All of the changes are more or less noticeable; the uneasiness, pain, bearing down, and watery discharges always accompanying the expelled ovum are usually mistaken for weakness, misplacement, leucorrhoea, etc., but by carefully noticing these disagreeable feelings it will be seen that they occur periodically, and are accompanied by peculiar mental conditions. At one stage a strange nervousness is manifested, at another, an unaccountable irritability of temper appears, quite unlike the natural disposition of the individual, then comes an inexpressible sensation of loneliness, often followed by a degree of languor difficult to account for. All of these symptoms do not appear in one individual, sometimes one or two; but occasionally all are present.

In this department of science, as in all others, a little study and observation puts one in possession of these facts, by which the number of offspring may be controlled, as well as the determining at which season they shall come; for there is a proper and an improper time for the young to come. The stock breeders understand this fact thoroughly. Animals are not permitted to breed at haphazard in mid-summer, mid-winter, or autumn. It is well known that the animals born at these unseasonable times are never so strong as those born in the spring. The earth is controlled by positive and negative seasons, as from twelve at night till twelve at noon we have a positive or magnetic season, and from twelve at noon till twelve at night the negative or electric, as from Winter Solstice to Summer Solstice we have a magnetic or positive period, and from mid-summer to mid-winter a negative. During the positive season there is taking place in all the ramifications of Nature a crystallization of forces. Nature is a unit. Therefore man comes under and should be controlled by these periodical laws. The best progeny, as a rule, is that coming in the early spring.

Statistics demonstrate that the rate of mortality is a third higher among children born during the last five months of the year, as this brings teething, that most trying period, during the latter portion of the heated term.

What the world stands most in need of to-day, so far as the progeny of the human family is concerned, is a *better* quality, and not a *larger* quantity. No family—unless the physical and pecuniary resources are un-bounded—shall exceed four children, and no couple desiring health and permanent happiness should be long without one or two at least, always providing that the physical and mental conditions, as well as the ages, are such as would ensure healthy and well balanced offspring.

Every child should be so organized, mentally and physically, that as it reaches maturity it may be enabled to advance the world's work, instead of retarding it through mental or physical incapacity.

When children are desired, a thorough preparation on the part of both parents is a duty which they owe to humanity; a preparation and a consecration of both body and soul. Particularly is this true where the balance in the physique has been lost through excessive labor or disease, or where the mental equilibrium has been disturbed by abnormal excitement.

There should be an abstinence from sexual indulgence for a considerable length of time, that all the forces may be employed in building up the system.

The farmer carefully prepares and enriches his soil before committing to its keeping the grain, which he selects with utmost caution, understanding that it must be fully ripened and of uniform growth to insure a crop which will repay him for his time and labor. Even the stupidest of the class understands that inferior half-ripe grain, planted in cold or worn out land, will yield nothing but disappointment to the laborer.

This disregard for a suitable preparation and proper selection of temperaments in the parents gives rise to the deterioration of the physical **and** mental qualifications so noticeable in many families.

The stock breeder understands that in this precaution lies the great secret of successful stock-raising. These are the points to be most considered in the selection of partners, particularly when the well-being of the progeny is regarded.

But unfortunately no attention is paid to this important law by people entering into that most sacred of all relations. Marriage is pronounced a lottery, **and so it** proves in most instances, **because it is** entered into blindly, without any study **in the** matter, regardless of temperamental adaptation **or mental** relation **to each** other. **The** considerations are usually wealth, position, or lower **still, mere** animal gratification.

INFANTICIDE AND ABORTION.

Much **has** been written, and still more said, concerning the sin of infanticide and abortion, and the strong tide of public opinion against this crime **has** proven a powerful check to its increase, **or** at least to the shameless brazenry with which the "slaughter **of the** innocents" has been carried on.

Abortion, the dislodgement of the embryon at an **early** period, **is** still practiced by many good, though thoughtless women, **who do** not consider it a sin, beyond the injury done their physical system, believing there is no independent after-existence for the foetus until after quickening, considering that so long as only the embryo is expelled, it is no more than the loss of an ovum. These women are perfectly conscientious in this belief, and consequently when maternity is forced upon them, through a hated contact perhaps, they resort to what they believe to be only a physical wrong toward themselves; and from our intimate knowledge of women as a whole, we are led to confidently believe that they only require a more extended knowledge on this subject in order to arouse their conscience, and impress this matter upon their minds in its true moral aspect. If they were fully convinced that this wrong, although unseen by man, was known and disapproved of God and angels, and lived as long as the soul lived, it would do more toward checking this sin than all of the penalties enforced by man-made laws.

If they can be made to comprehend the fact that the soul is the nucleus of the new being, and the body merely its outgrowth, that after the generative act has been consummated and impregnation has taken place, that a soul has found lodgment here and must ever after have an independent life, thrown adrift upon the vast ocean of existence to live its life and develop its resources as best it may, after the manner in which waifs are deserted and cast adrift in this life.

If this view, which all nature confirms, was presented instead of legal penalties, the gigantic crime of infanticide and abortion would be rarely known, but so long as men, through passion alone, impose the burden of maternity unasked and undesired upon women, and until women are educated to a higher sense of their responsibilities, this, and kindred wrongs will continue to be committed.

Women must control these matters themselves—must determine when the generative act shall take place. We repeat, they must determine this for themselves! The woman who submits to the embrace of a man against her will and whom she dislikes, is living in one sense a life of *prostitution*, although bound by the strongest ties of legal marriage. Such a life is detrimental to both soul and body, and largely the cause of this wholesale murder. The woman who loves the man by whom she becomes pregnant will hold and cherish her babe with the same love which she bears the father, and would no sooner destroy or harm this tender life given into her keeping, than she would do violence to the author of its being.

There can be but little doubt but what repeated abor-

tions produce serious uterine diseases, such as induration, ulceration, and cancer; and every form of misplacement from weakness, to say nothing of the shock
to the general system. It is the greatest physical outrage that can be perpetrated upon a woman. The
writer has known of nine cases of partial paralysis of
the lower limbs from this wrong being several times
repeated, and a great number of cases where life has
been sacrificed through inflammation of the uterus,
frequently miscalled inflammation of the bowels.

These accidental impregnations must be avoided.
The sexual question is the vital one of the day. It
looms up a gigantic barrier in the path of progress.
Who shall solve it?

Upon woman alone devolves the safe adjustment of
this matter. She has the right of control over her
person at all times. It will require time, patience,
and womanly tact to convert man to this view, but the
benefits, physical and mental, accruing to himself from
this abstinence will in time fully convince him of its
wisdom and justice.

CHAPTER IX.

HEREDITY.

The law of heredity should be thoroughly under-stood by every man and woman entering into the mar-riage relation. This **knowledge** possesses a double importance in the mother's **case, she** having a two-fold **work** to perform. It would **enable her not only** to gov-ern her life during gestation more judiciously, but to better understand the mental **and physical** peculiarities of **her** children **and** thus be prepared to train and develop them according to their individual needs, as no two children should receive, in all respects, the same training.

The masculine element of character, is strength; the feminine element, harmony, and through its refin-ing influence the mother is enabled to subdue the strongly marked eccentricities transmitted by the father, and counterbalance, to a degree, evil tendencies of a vicious and depraved sire; as in the case of Mad-ame Maintenon, who inherited her varied talents from her unfortunate and infamous father, but the harmony of her character from her devoted and affectionate mother.

But, unfortunately, all women do not possess this refining characteristic, **and** when women **not thus** endowed become mothers, **we may** reasonably expect to **see** weakness, idiosyncracies **and** angularities, even though **the** fathers may **be well balauced and** superior men.

History tells **us that** the **mother of Christina of** Sweden **was a woman of weak judgment and** capricious temper, **but** that **the father of** this remarkable **and** eccentric **woman was** one of **the** greatest and noblest men known to history.

Erratic Madame **Krudener is said to** have been **a perfect** counterpart of her father, mentally and physically, **her** mother being **a woman of** inferior intellectual capacities **and** entirely **destitute of the power** to discipline **her family properly, which no** doubt influenced her daughter's diversified and brilliant career.

When there is a correct cross in the temperaments of the parents, we find, as **a** rule, the daughter inheriting the mental and physical characteristics of the father, and at times the brain **of a** sculptor, painter, philosopher, metaphysician, scientist, is contained in the cranium **of a woman;** this **brain** requiring the same food and stimulus for proper growth and expansion as

33

that of the father's, and frequently united to this mental conformation we have a robust physique which gives increased activity to all of the powers. A careful study of biographical history will prove, with few exceptions, that the most talented women of all ages have been the daughters of men of great intellect and strong will.

Under the same law, boys, with few exceptions, represent the mental, moral, and physical conditions of the mothers, and men who have wielded the widest influence for good to the world of mankind have been the sons of women possessing clear, well-defined mental capacity, great moral strength as well as the feminine quality of character.

Just how this exchange of temperaments, talents and peculiarities from fathers to daughters, and from mothers to sons is effected, would be difficult to tell. We only know that the facts exist and should be regarded, in order to adjust the surroundings and give to each soul an opportunity for the unfoldment of its individual capacities.

This subject presents another important consideration, and that is, that the mother should be of the best possible type of womanhood in order to give the world more perfect specimens of manhood.

In contemplating the life and writings of Victor

Cousin, one can only think of him as the son of Madame Cousin, but never as that of the humble artisan, the commonest of his kind, who saw nothing better in life for his boy than to become like himself—a watchmaker. The early neglected, though gifted mother felt keenly through her own unsatisfied craving, the needs of her child, and solely through her efforts has the world been blessed with one of its most advanced thinkers.

Said Benjamin West: "My mother's kiss made me a painter." That kiss may have aroused his artistic genius, but did not make him an artist. The painter slumbered in the soul of the mother long before the boy had an existence.

Cowper inherited his poetic talent from his mother. She was, it is said, a gifted and beautiful woman. Cowper so closely resembled her in personal appearance that his cousin, Lady Hesketh, used to playfully part his hair in the middle and declare that she could well imagine that her aunt Cowper sat before her, so striking was the likeness. Between the unfortunate poet and his mother there seemed to be a chord of sympathy which death could not sever; for although he was quite young when she died, still she was his guiding star through his after life. Talking once with his tried and valued friend, Lady Throckmorton, of his early life, he said : "Although nearly fifty years have passed since my mother's death, yet not a week passes, I may truly say, not a day, in which I do not think of her, such was the impression left upon me of her great tenderness, yet strength of character."

Dante derived from his gifted young mother his wonderful poetic genius, as well as that peculiar spirit illumination so strongly manifested in his "Inferno." Dr. Davis tells of thrilling visions seen by the mother, shadowing forth the future greatness of her son.

Goethe says of his parents : "From my father I derive my frame and the steady guidance of my life, and from my dear mother my happy disposition, and love of story telling." His biographer relates that Goethe inherited from his mother that large and instinctive wisdom which comes of broad human sympathies. His mother had a great love of poetry and romance, a sunny disposition, and withal was a great philosopher. Goethe's father was a stern man and a rigid disciplinarian, and doubtless to this fact the son owed his scholarly attainments, while from his mother he derived his genius.

George Washington closely resembled his mother, both in his physical conformation and mental endowments. History says that she was a remarkably intelligent and energetic woman, who ruled her family

and household with a strong hand and firm will; and
that she not unfrequently shared the labors of the field
while directing her servants. Washington was only ten
years old when his father died. He was often heard to
say that he knew little of him; that to his mother's
care and counsel alone he owed his success in life.

The father of Napoleon does not occupy a prominent
place in history, while his mother stands out pre-eminent, exhibiting through life many traits of character
in common with the illustrious diplomatist and soldier.
For weeks before his birth she rode on horseback beside
her husband during his fatiguing marches, and at the
close of the Corsican campaign hastened to Ajacco, her
home, having only been there a few days when Napoleon
was born. The circumstances of his birth are quite as
interesting as are those attending the more thrilling
episodes of his maturer years. Madame Bonaparte was
attending mass when she was seized with the pains of
child-birth. She left the church, hastened to her home,
and had only time to enter the great hall, hurriedly
throw a robe upon the floor and place herself upon it,
when the renowned hero entered the world. These
circumstances display a bravery and determination of
character in the mother which must have influenced
the life of the son. Napoleon entertained for his

mother an unbounded reverence, attributing to her the promotion of the entire Bonaparte family.

The following examples of the transmission of talents and peculiarities from fathers to daughters are equally marked and noteworthy. Catharine Macauley, the historian and politician, may be cited as an illustration proving this theory. Whilst Catharine was yet an infant, her mother, a quiet, unassuming woman, died, leaving the babe to the care of the father and a governess. At a very early age the precocious child showed a fondness for books, her father giving her free access to his large and fine library, permitting her to make her own choice in selecting her reading matter. After a time he directed her attention to those studies which had been most congenial to himself—History and the Science of Government. Here he touched the master chord of her strong, deep nature, and gave to her mind an impulse in the direction of politics and historical research.

The remarkable intellectual and physical development of Madame Necker was inherited from her father, who was not only a man of magnificent presence, great mental powers, and genuine kindness of heart, but also very learned. He trained this lovely and talented daughter with great care, giving her the **severe** and classical education **usually bestowed** upon men only.

The **famous Margaret** Mercer so closely resembled her father in every particular **of** character, that **he** resolved **to give her, as far as** possible, the benefit of **his own** education, which **was a very** superior one. This resolution he carried out, Margaret completing her course of studies entirely under **his** tuition.

Dorothea of **Russia did not exhibit a trait in** common with **her mother, but showed in** an eminent **degree the** powers of **her** accomplished **and gifted father.**

Queen Elizabeth presents **in her character none of** the grace and beauty **of** Anne Boleyn, but instead, the **blunt,** rough sturdiness **and** dogmatism of Henry.

Bishop Burnet says of Lady Falconberg — Oliver Cromwell's third daughter—that **she was better calcu-lated to** have **maintained the** post **of** Protector **than** either **of her** brothers, **who were in every particular** the counterpart **of their mother; she being an ordinary** woman **with** neither, **wit,** beauty, nor education. It is **a** well known fact that Lady Falconberg exerted great **power** and influence, **and** that she largely contributed toward **the** Restoration, possessing strong mental traits, peculiarly like those **of** her **father.**

Harriet Hosmer and Rosa Bonheur both strongly

resemble their fathers, even to a degree of masculinity, and instead of being educated in the usual conventional manner of young girls of the day, were permitted to choose their avocations, following the direction of their strongest and best faculties, and have thus incalculably enriched the world of art.

Had the father of Sirani persisted in keeping her to the harp and embroidery frame, the **usual** employment **of ladies of** her **rank, Italy,** and we may say, the world, would have lost one **of its** greatest artists.

Giacomo **saw** much of **his own** power in his daughter Marietta, and instead **of educating** her for the usually restricted sphere of the women of **her time, he** carefully developed her wonderful genius **in the line of** art, and the result **was one of** the best portrait **painters** of all Italy.

The father **of** Angelica Kauffman once said: "I should never have thought of instructing my daughter in the art **of** painting, but for the **fact that** when quite an infant she would **so** exactly imitate **me in** holding a brush, that it occurred **to me as** being possible that a girl **might be taught** the **art."** All the world knows the sequel.

Sarah Siddons **and** Charlotte Cushman present two striking illustrations of paternal **transmission.**

Every woman who has made a success of her life through her own exertions has been, as a rule, her "father's own girl."

Children rise above, or sink below, the paternal standard in proportion to the exchange of temperaments between father and daughter, mother and son.

For example, it has almost resolved itself into a maxim, that a girl born the exact counterpart of her mother, physically and mentally, is born to ill-luck, and a boy resembling his father in every respect will not be his equal in calibre, and when this exact likeness between the male members of a family is carried into the third generation, there is usually manifested both mental and physical deterioration. This is often the case when the grandfather has been a man of ability.

There is an old and trite adage which expresses a vast deal of philosophy in a few words, "Boys are like vinegar, the more mother they have in them the sharper they are."

We hear much among a certain class about "lucky" and "unlucky" people. Good or bad luck, as the saying goes, is no doubt born with the individual; that is, the faculty to advance or retrograde is inherent, and inasmuch as every man and woman fashions his or her destiny through will and energy, or the lack of these qualities, they may be said to be "lucky" or "unlucky," as the case may be. But in the truest sense in this square-headed, reasoning age, luck signifies *pluck*, energy, foresight, the power to reason from cause to effect—to see the end from the beginning—with a determination to meet all emergencies, and by physical energy and will to overcome all obstacles.

CHAPTER X.

MOTHER'S INFLUENCE.

THE French proverb, "The world is woman's book," is a most significant one, the truth of which is but little appreciated by womankind at large. Women are disposed to depreciate their influence over the lives of those with whom they come in daily contact, over fathers, brothers, husbands and children. No circumstance, however trivial, but has its influence over the character of a child. From the age of two to that of sixteen, the young of both sexes are momentarily receiving impressions. This is the period during which the world, in truth, becomes woman's book.

Every decided action, every emphatic thought expressed, is recorded therein, moulding the mental, moral, and religious characters of the boys and girls who make the men and women of the future.

Boys, especially, are more easily influenced by their mother than by their father, owing to an attraction resulting from a difference in the sex, she having the power to create in them the most exalted ambition, and develop those noble qualities which constitute true greatness.

Napoleon frequently declared that his family were solely indebted to their mother for that wholesome physical, intellectual and moral training which prepared them not only to ascend to high positions, but to maintain them with dignity after they were attained.

The mother of Cuvier early took the charge of her son's education, counselled him to read only the best books, selected herself the works on literature and history, creating in him that thirst for knowledge for which he became so famous in after life.

Not only will kindness, tact, and patience be required, but a knowledge of the individual characteristics of the children.

The father of John Wesley once said in a conversation with his wife concerning their eccentric and gifted boy:

"I have heard you tell Jack twenty times to do that one thing."

"Had I been satisfied to have told him only nineteen times, I would have lost the entire result of my labor," was the patient and womanly answer.

Richter says that the boy of fourteen is on the boundary line between the monkey and the man, and at the most trying period of his life. It is indeed the turning point at which he is made or marred.

Lord Byron attributed the failures of his life to his mother's unhappy influence over him in his youth. Her frightful paroxysms of anger, want of sympathy, and heartlessness in taunting him with his deformity embittered his whole existence and destroyed his confidence in both women and men. He says of her: "She almost drove me frantic daily by her taunts and insults." It is said that he accredited his domestic troubles largely to her unhappy influence upon his character.

The same moulding influence which gave to the world a Nero, a Caligula, a Vitellus, gave also a Tasso, a Lamartine, and a Dante. Lamartine's mother was his inspiration. Those sublime flights of fancy and eloquence by which he so often swayed the public mind for good, are due to her early training and judicious selection of that class of studies which developed his poetic genius.

Plutarch in his writings, speaking of the mother of the Gracchi and the education she gave her children, says: "Cornelia brought them up with so much care that, though they were without dispute of the noblest family and the happiest geniuses of any Roman youth, yet education was allowed to have contributed more to their perfection than nature. This woman was left at the death of her husband with twelve children, and those of the number who survived, were wholly educated by her."

The mother of **Immanuel Kant,** when he was a mere boy, would hold long conversations with him upon the most abstruse metaphysical points, would question him as to the means by which he was enabled to think, thus leading his thoughts into those avenues that ultimately opened into such vast fields of speculation and research.

Hegle, the German philosopher, received up to his sixteenth year his entire education from his mother. She taught him the languages, and compelled him to make extracts from all the important works he read.

Thomas Chatterton's father died when he was three months old. At the age of six years he was dismissed from school, being considered by the teacher an imbecile. His mother now took charge of him and attracted his attention to poetry, of which she was passionately fond. At eleven years he began writing and at twelve had completed his "Elinore and Inge." The mother and sisters of this eccentric genius were the objects of the most intense love and devotion.

Bulwer Lytton was deprived of his father when he was very young, but he was fortunate in possessing a mother who had cultivated a decided taste for literature. She did much toward forming his mind, and it was to please her that he, when only six years of age, wrote his first verses, thus encouraging him to renewed efforts, thereby laying the foundation for future greatness.

An impassioned Vendean woman with a soul fired with devotion for the Royalists, whose cause she so boldly espoused as to lead to her proscription while yet a young girl, became the mother of Victor Hugo, and into his young life, day by day, in those hurried marches over the country, she infused her spirit of loyalty and fearless devotion to what seemed right and just.

Louis Blanc imbibed his zeal and patriotism from his enthusiastic and liberty-loving mother.

Gerald Massey, whose hymns of progress so stir the soul of all lovers of humanity, owes not only his genius but its cultivation to his mother. His father was a canal boatman, one of the poorest and most illiterate of his class, but his mother, although uneducated, possessed a finely organized brain and a keen sense of the importance of a more perfect mental development, and in every way encouraged her son in his efforts to obtain an education.

The mother's influence upon the daughter is no less marked and lasting in its effect upon the character for good or ill. The most evenly-balanced and harmoniously developed women are those who have been trained by intelligent and tender mothers. Understand, we do not **say** the most talented, but the most lovable, wise **and** womanly.

Not only does history furnish innumerable illustrations of this fact, but the truth is daily being demonstrated in the humble and unwritten histories **of** every-day life.

Queen Victoria presents an interesting example of the effect of a mother's influence. **Her** education, intellectual, moral and physical, was conducted entirely **by her** mother, and accomplished **almost** exclusively through female agencies. Her father died when she was but eight months old, **and to her** mother, the Duchess of **Kent, was** left **the** sole guardianship of the **royal infant,** and from that time until the Princess ascended the throne they were **never** separated. The Duchess nursed her babe at her own breast, and as soon as the child could sit alone she always dined with her. In the class-room **she was** still her companion, encouraging her by the lively interest and sympathy she manifested in her studies **and** amusements, never allowing the demands consequent upon her exalted rank to interfere with her duties to her child. Doubtless to the wise moulding influence of this mother England owes her peace and prosperity **to-day.** What the Duchess **of** Kent did **for** a nation, mothers in the lower walks of life may do for families. No character can even approach perfection that has not been formed under the fostering care of some good, sweet-souled, unselfish woman. It is a misfortune that almost amounts to a catastrophe for a young girl to be separated from the sympathy of her mother, or some older and more experience female friend.

For a mother to complain that her daughters are unfilial or disobedient, is to acknowledge that she has failed in her government. Our children, like the world at large, will take us at our own valuation. The mother who makes a slave of herself that she may keep her daughters in idleness and luxury will receive in return only neglect and contempt. The extravagance, frivolity and idleness of the majority of our young girls have become proverbial. Instead of being trained to habits of industry they are carefully shielded from all care and responsibility, and as a matter of course are wholly undeveloped in body and mind. When we say undeveloped, that scarcely expresses it, it is even worse than that. As there is no such thing as standing still, therefore, if the young mind is not occupied in the direction of good, it will be in the direction of wrong.

A mother has a mighty responsibility in the care and training of her children.

"A stitch in time saves nine," is a good old maxim,

35

applicable at all times and in all departments of life, and nowhere more so than in the training of children, particularly daughters. **If** mothers would establish **a bond of** sympathy between themselves and their girls, **the power to** influence them for **good** would be a comparatively easy matter. **Every woman** who respects and loves her mother, respects **and** loves **all** womankind. She who has no regard **for her** mother, **has** none for sister or friend. A confidence in, and love for her own **sex** should **be** early instilled into the mind **of** every young girl, **for notwithstanding the** prevailing idea **to** the contrary, based upon superficial observation, woman **is** woman's best friend, and the woman **who** chooses her **most** confidential friends from **the** masculine ranks instead of **from that of** her own **sex has** made a mistake, **not** only evolving serious difficulties, but one which will deprive her of such sympathy and counsel as only a woman can give. Woman represents the best side of humanity, so created because she was to become mother of the race, and being **created** last, **was** an improvement **on** the failures in **man.** According to Burns—

> " **His 'prentice** hand He tried on man,
> **And** then He made the lassies."

It is **a** matter **of** congratulation **that** His Satanic Majesty **is of the** masculine persuasion ; that poor Eve did not, after all, plan the downfall of the race.

CHAPTER XI.

WILL POWER.

It seems impossible to over-estimate the power of the individual will, over the mental and bodily functions, when we take into consideration its influence in general upon the affairs of men. It is the chief engineer that carries out the plans of all the councils; the various faculties of the mind consult, advise, devise means and ways, and after mature deliberation, having no power in themselves to act, send in the order for the will to execute. It is wisely said that "Will, not talent, governs the world." It bridges rivers, chasms, and bayous; it girds the earth with railroads and telegraphs; it tunnels mountains and lakes; it covers the oceans with ships of commerce; it cleaves the earth for hidden treasures; it founds colonies, builds cities, changes deserts into gardens; gives meat and drink to famishing thousands; and over the mental and physical domains in the individual, it can be brought to act quite as effectually. It will ward off disease, and, at times, even death is vanquished by its power.

Like all the other faculties, it can be cultivated; 'tis a creature of growth, and forms the real basis of character. If it be strong, then the character is likewise strongly marked. We meet men and women daily whose intellectual endowments are of the first order, yet who accomplish nothing in life, but are drifted hither and thither at the mercy of stronger wills, having not only their own sorrows to bear, but the miseries and despondencies of those about them. Such people are more subject to diseases of various kinds; especially is this true in the cases of sympathetic, sensitive women; in a large majority of cases their maladies growing out of mental worriment and despondency, which adversely affects the bodily functions.

Mrs. Frank, our next door neighbor, a dear, kind, sympathetic soul, good enough to be an angel, but not quite strong enough to be a human, comes to me every now and then to ask what she shall do in order to ward off the unpleasant outside influences that so try and vex the soul. This morning she came, full of trouble about our new neighbor, Mrs. Stately, who lives on the other side of the street.

"I am actually afraid of Mrs. S," said she, in her frank, honest way. "I can't endure to have her touch me; I cannot tell you how I feel about her, yet every one says she is such a nice person. Now is it my fault? You know I wouldn't do her a wrong for the world, but every time she comes to our house she is sure to make unkind remarks about some one, and that stirs me all up; then after she has gone, I have a good cry, and the headache all the next day. I didn't sleep one hour last night, just from the effects of one of these visits. I do wish that people would stop talking to me about such things; why, Mrs. Stately told me the most scandalous story about a dear friend of mine, a story that I know to be utterly false. Now, what can I do?"

"Why didn't you knock her down?" I suggested, in a mild way.

"Now, don't you be ridiculous," and over-wrought Nature could bear no more, and the poor little woman cried as though her heart would break.

After a moment's pause, I asked: "What did you say to Mrs. Stately after she told you the story about your friend?"

"Why, what could I say? I did not wish to offend her; she is a woman of great influence, and I do not like to have any unpleasantness with my neighbors. Then, too, she told it with such an air of assurance as though no one would dare to dispute her authority. How I do hate such people!"

Then she fell to crying again. Sitting down beside her and taking her hand, I said :

"This morning I heard you telling the children when they went out to play, to be sure and not go near that three-leaved ivy which grows over the wall at the end of the lawn, saying that you had once been badly poisoned with it yourself, and that John, the gardener, was always obliged to wear gloves when he repaired, or in any way worked about the wall. Now there is so much that is good and true and beautiful in this world that seems hardly worth our while to cultivate what is ugly, unwholesome or poisonous. But if they will grow in spite of us, we must do as John does—*handle them with gloves*, or avoid them entirely. It does not pay to fret over those unpleasant obstacles which cannot be removed. It would be most unwise to complain because winter came, but most wise to protect the tender body against fierce blasts."

"But how can we escape these things ?" asked our neighbor anxiously. "I am really sick after seeing or hearing of any suffering or serious trouble, even though I may not be at all acquainted with the parties in question. And when people come and tell me of any scandal or gossip about my friends I am wretched for days."

"Is any one benefitted by these days of wretchedness?" I asked. "The individual who allows his sympathy to run away with his judgment can be of little use to any one. 'He does but little who gives his tears.' What good can you do these friends in the condition that you are in now? What can you do for yourself or family? We want backbone in our friends; tears don't fight our battles. Cease crying at every little thing. Cultivate your will! Use your bicep muscles more; and your lachrymal glands less; get more backbone. Learn to say, I will, and, I will not. Build about yourself a wall behind which to protect yourself from these assaults; in a word, ensphere yourself in your will! Double up your fists; straighten up and stand on the defensive for a time, and see how soon these headaches will disappear."

"Tell me how I can do this," cried the eager, excited visitor, drawing closer, and looking up in my face with her earnest brown eyes.

"Dear heart," I said, "I have once suffered all that you are now suffering, but have long since learned not to cry over the ills of life, and when I cannot put them aside I just step over, or go around them. Years ago I suddenly awoke to the fact that the world at large did not respect me the more because I made myself miserable, and I also found that crying brought wrinkles, and lastly, it came to me like a revelation that for all of my unhappiness, tears, repinings and wrinkles, the world was not one whit the better, but I instead, sadly worsted in the conflict. I gathered up my scattered forces and sought nature for consolation, and to my interrogation she gave me this philosophy which has been to me a religion.

"All outward forms are simply envelopes with which spirit clothes itself. The life of the egg is shielded outwardly by the coarse shell, then by a membrane, then by still more delicate ones, fold upon fold, in order to ward off injuries and keep intact its life. The tree presents its rough bark, whilst shrined within is all that makes bud, blossom, and fruit. If portions of the outer coat be removed, it will be at once restored, else the tree, which is only the form which the inner life wears, will be sacrificed. Those forms which are to be employed to express the demands of the soul are more perfectly protected in all directions. In Man, he being the highest type of animal life, Spirit has encased itself in a complicated, yet harmoniously organized body. The brain, the soul's most important instrument, is encased in a sphere of hard bone, and, like the egg, has its numerous and elastic membranes. The spinal cord is, like the brain, enveloped in its appropriate membranes, and surrounded by its sharply spired bony rings. The eye has its membraneous

envelopes and aqueous fluids which shield the inner and
real organ of vision. All organs of the body are thus
protected; and over this complicated soul garment is
an insensible coat—the skin, and over that we place
other envelopes, in form of various garments, and still,
above all and over all of these guards, are erected
houses with double and treble walls.

"In these manifestations may be seen that self-pro-
tection is the ruling instinct of animate life. This
is seen in the lowest grades of organic life; in the coats
of the most minute cells that go to make up living
organisms. But all of these guards and shields are
merely intended to resist those evils and accidents
incidental to matter, such as tempests, floods, accidental shocks, extreme cold and heat, but not to
protect the soul. They tend, it is true, to keep the
body intact, thus affording the soul a habitation; but
the real life is the unseen life, which cannot be protected by mere bones, membranes, clothing, houses,—
indeed, by anything which can be felt or seen. No
material guise could protect a sensitive soul from the
shafts and wiles of a vicious one.

"Throughout the visible world we have seen that spirit
life is shrined and guarded by fold upon fold of matter,
sphere within sphere, cell within cell. Analogy teaches
us that the soul of man is thus enshrined within a

spiritual and mental **sphere.** **It is** wholly through the Will that the body is clothed, fed, and sheltered, making **use of** such material substances as are suited to its needs. In a like manner, the Will fashions from the subtle spirit elements a guise for the soul to wear.

"A weak will meagrely supplies the physical needs. You say that John has not life enough to stoop down and pick up a ten-dollar gold piece if it lay in his path. John is almost destitute of **will.** Mark his slovenly gait; **see the** careless **arrangement of** his dress; **his soul is** just **as poorly** clothed **as** his body. **It is exposed and racked** by the temptations, fears, wrongs **and sorrows of** every other kindred **soul** with which he comes in contact. Such persons laugh with those **who laugh, and** weep with those **who weep.** The spirit atmospheres **of** the malignant and vicious penetrate through the thin garments which the weak will has thrown over such defenseless souls; and they are made to suffer as the body would, when insufficiently clothed and exposed to rigorous and tempestuous weather.

"The **power** to shield yourself **from** the annoyances of which **you complain is** under your own **control.** All you require, is to cultivate your Will Power and **reason,** silencing **all persons** who would annoy you with gossip and scandal. And above all, put under your feet those burdens which were never intended for you to carry, and remember that gloom never makes the world brighter or happier. Sunshine, not clouds, brings flowers and fruitage."

My little friend, smiling through her tears, said: "But I am such a weak little thing, what can I do toward making these people better or kindlier?"

"Don't depreciate your influence, dear heart," I urged. "The world is made up of small things. A straw may turn the current of a mighty river; from the infinitesimal zoöphyte grew the coral reefs and islands. Your little words of admonition may make them pause and like the straw, turn the current of their lives from their present course into deeper and truer channels. Every woman was designed for an educator. She may not paint as Raphæl did, write like Homer, carve like Phidias, or design like Michael Angelo, but if she be a true and earnest woman, seeking only the highest welfare of others, her life will be incalculable in its good results to humanity.

"Richter says: 'All the energies with which nations have labored and signalized themselves, once existed in the hand of the educator.'

"As the acorn holds within its tiny cup the giant oak, so you have within your nature undreamed of

possibilities. You only need judicious mental gymnastics to make you intellectually and spiritually
robust.''

Just here Willie Frank came to tell mamma, that a
lady wished to see her, and rising, my visitor departed,
turning toward me, as she passed out, a smiling, earnest
face, which gave me courage to believe that she would
be ready to meet the emergencies of the morning,
whatever they were to be.

CHAPTER XII.

UNSEEN INFLUENCES.

THAT Mysterious Power that called into being the innumerable suns and their satellites that now roll in space, strung them upon a continuous chain of gradations, the links of which enable the careful student of nature to feel his way from the lowest strata of protoplastic life through consecutive stratas to the most sublimated essence of things.

There is no break in this chain. On our globe, we observe the wonderful interblending of the several kingdoms; mineral life merging into vegetable—vegetable into animal—animal into soul—soul blending with that mysterious Infinitude which man instinctively reaches out for but dimly comprehends; and pervading and uniting all is a subtle spirit essence—unseen, but most potent in its power. It pierces matter to its lowest depths. It forms a means of subtle communication from plant to plant, and from animal to animal, and from man to man. He, being the epitome of creation, is in close affinity with all that exists in the lower kingdoms, all things yielding up their spirit to him, he unconsciously renewing his physical life through their ministry.

This delicate media of exchange between man and man not only enables him to read correctly the secret motives of others, but to an extent interpret the past lives of those with whom he comes in contact. This is man's safeguard; he need never be deceived, if he is a careful student in character reading. To admit that he has been, is to acknowledge himself a dullard in the great school of observation.

Through the "soul of things" ,we reach the secrets of life.

Standing upon a table in the distant corner of the room in which I am writing, is a bouquet of freshly gathered flowers. I have not examined it, yet I know it is composed of roses, carnations, mignonette, tuberoses, etc. How do I know this, do you ask? In this way: Through thousands of delicate nerve filaments I am absorbing into the more refined portions of my being their spiritualized life. Each has an aura peculiarly its own by which its presence is made known. This is in obedience to the law governing the "soul of things." Every form of animal or vegetable life tells its own story. The rose gives out an unmistakable odor by which we are made to feel its presence; the stately pine, that emblem of constancy, breathes out life to the exhausted nerves and diseased lungs; the grand old ocean gives out ozone which renews the wasted energies of the blood.

The sun is millions of miles from our planet, and, according to the calculations of an eminent philosopher, there rolls between that stupendous globe and our own, oceans of ice, yet the flesh, brain, and blood, are pierced by innumerable life-giving emanations that pass off from that great source of light and heat. We cannot see these subtle agents, yet we know that they exist. Electricity and magnetism elude ocular detection, yet how powerful is their influence for good to the race. These are the tender nurses employed by Mother Nature to restore her children to health, and to keep them within the reach of her great, loving arms.

The mental, spiritual, and physical emanations from the human species are as unmistakable and as invisible. In the souls of many women we recognize the modest violet with its sweet perfume; in others, the mignonnette, with its plain dress and spicy breath—roses, with their delicious odor—stately pines, with their aromatic spirit, imparting life and strength—great, ocean souls, in which the over-burdened may bathe and find rest, for a moment's clasp of a healthy hand will leave in the palm of fevered discontent, an influence which may cool the pulse for days.

Evil exhales its spirit as unmistakably as does good. The noxious henbane and nightshade warn us of their presence; the vast marshes festering under the burning sun throw out their infinitesimal death-dealing needles of alternate ice and fire, which pierce the quivering body through and through.

There have been periods in the history of Rome when it was almost certain death for a stranger to enter her gates. Yet, far or near, nothing met the eye save the glory of Art and Nature combined; but outside of her walls, unseen by the luxurious dwellers upon her seven hills, extended malignant marshes, with their deadly exhalations, which grasped with their invisible fingers of death, the vitals of the luckless stranger.

On the dark side of human life we find souls that resemble the deadly nightshade, the poisonous henbane, miasmatic marshes, with their pestiferous breath; we need not be told their character or their proximity, for every woman, who in passing you touches your garments, lays her hand upon your arm, clasps your hand in friendly greeting, has left some portion of her being with you, by which you may read her correctly, and in return you have imparted something to her. It may be the exquisite perfume of the violet, the spicy breath of the mignonette, the cool, balmy, ocean breath : or it may be the sickening exhalation from a diseased and discordant soul.

The law of economy is so rigidly observed in the government of the universe, that nothing is lost. Not a thought emanating from the human brain, but lives somewhere. Those with whom we come in contact read our secrets; our thought emanation settles upon them as dust from the wings of a miller. No life can be lost, or even hidden. Murderers, liars, and thieves carry about with them the proofs of their guilt.

We have our mental, as well as our physical atmosphere. The impure thoughts of strongly magnetic persons will influence, without a verbal interchange, the thoughts of those who are more susceptible and less positive than themselves. Our evil thoughts may be productive of evil deeds in another, although we may never have shaped them into words or deeds. Wrongs and crimes, done in secret and darkness, walk abroad like spectres, in open day; known of God and seen of men.

A curious incident occurred some years since in the family of an intimate friend, whose name, from motives of delicacy I withhold, as also the locality in which the occurrence took place. I will simply relate the facts as they transpired, leaving the reader to draw his or her own conclusions regarding them.

37

This friend, having a family of four small children, had thought it advisable to go to the country during the heated term; and in order to give greater freedom to the little folks, it was deemed expedient to keep house rather than board. At length, from a list of advertised houses one was selected as being every way desirable, and all preliminary arrangements being made, the family took possession of their new home. In the course of a few days a gossipy neighbor called, and among other items of interest, informed my friend that the house was formerly owned and occupied by a clergyman, who had, in a moment of mental aberration it was supposed, committed suicide by hanging himself from the knob of his study door, and since that time no one could be induced to live in the house. The lady was considerably shocked by the information, more especially as she had converted the room in which had been enacted the terrible tragedy into a sort of nursery and family room. But being a woman of sound judgment, she resolved to keep the matter entirely from her family, particularly from the servants and children.

A week passed pleasantly away in their new home, when an event took place that inexpressibly shocked the whole family, and left in the mind of the mother strange fancies ever after concerning unseen influences.

One morning, on returning from her accustomed drive, she was greatly alarmed as the carriage drew up to the gate, to hear confused noises, denoting some unwonted excitement in the house. Fearing some serious accident, she sprang from the carriage and hastened into the hall, where she found the servants, pale and excited, vigorously rubbing, shaking, and blowing into the face of the youngest boy, who presented every appearance of partial recovery from suffocation. When the child was out of danger, and the excitement had subsided, the nurse gave the following information relative to the affair.

Some time after her mistress had left the house, she, in passing through the hall had discovered that the nursery door was closed—a very unusual thing—which at once aroused her curiosity. She attempted to open it but was prevented from doing so by something from within. Upon more forcibly pushing it, however, it yielded, permitting her to enter. Here, strange to relate, she found the child with a rope about his neck, hanging from the door-knob, the *identical* one from which, three years before, the unfortunate clergyman had taken his final leave of the trials and cares of this life. With the characteristic coolness of a Scotch woman, the nurse had lifted the child up, loosened the cord about his neck, laid him on his back, and was using proper restoratives when his mother arrived. The most rigid examination failed to furnish any clue to the strange act of the child, as not one of the children or servants had ever heard of the tragedy.

Napoleon caused a sentry box in Paris to be torn down in which three sentinels in succession had hanged themselves.

Spanning a river in one of the Western States is a bridge from which three members of a family, one after the other, have leaped to their death.

It is not alone through the cunning of detectives that murderers are ferreted out. The circumstances of their daily lives cling to them as a garment, and proclaim to those coming in contact with them their real character. The cunning brain may frame lies for the lips to utter, but the mere words deceive only the superficial. Even animals possess the power to distinguish the approach of evil. We have reason to believe that the well-trained of the more intelligent species of dogs at times recognize the evil intentions of bad men. In fact there are incidents on record that clearly prove it. How much greater the need of man's being able to perceive this hidden intent on the part of another. The woman who meets you with smiles and overwhelms you with caresses, yet in her heart envies and dislikes you, will manage to leave a sting that will sometimes rankle for weeks.

In those benighted days when a belief in witchcraft held sway over the minds of men, the opinion prevailed that all persons who were in league with the Evil One possessed the power of torturing at will those whom they hated, by sticking their bodies full of invisible pins. The witchcraft period with its bloodshed and horror has passed away, but the power to torture souls by invisible means did not disappear with it. Only yesterday a lady paid me a visit, greeting me upon entering the room with kisses and much hand-shaking, expressing the greatest delight at the meeting. But before she had taken her departure she had stuck me full of imaginary pins, giving the keenest suffering for hours. I knew she would do this when I saw her enter just as well as I knew that there were tuberoses and heliotrope in the bouquet.

It is no more possible to analyze critically this subtle power, than it would be to describe the structure of electricity, or the size or texture of a thought. Every physical organization is an electric battery, possessing in health all the necessary chemicals for the production of a refined electricity.

Man is triune in his nature; he is both electric and magnetic. But more powerful than these, is his psychic power—a potent and far-reaching medium of which man, practically, knows but little.

CHAPTER XIII.

PHYSICAL PERFECTION.

COUSIN says "Physical beauty serves as an envelope for spiritual and mental beauty." This sentiment is in the main true, whatever cynics may say to the contrary.

Physical beauty is not so much the gift of inheritance as the result of cultivation, an outgrowth of a symmetrical soul, and does not consist in beauty of hair, eyes, and features merely, but in the *tout ensemble*. An erect carriage and graceful bearing sways a greater influence than any amount of facial beauty alone. A woman with a bow back and stooping shoulders would never be pronounced a beauty, although she might have the face of a St. Cecilia. A distorted body overshadows all other marks of beauty, from the fact that power to recognize form and proportion is the strongest of all of the perceptive faculties in man. The shape is the first to attract and hold the attention; if the bearing be fine, imposing, or graceful, the features not greatly out of proportion, the first impression would be one of beauty and power, and first impressions are usually most lasting.

Next in importance among the perceptive faculties is that of recognizing **color.** A clear, clean, and **smooth** complexion **at once** attracts the beauty-loving **beholder.** One need **not** possess a **white** skin in order to **be** beautiful, but it must be smooth **and** healthy. Owing **to the** fact that the skin on the face is constantly exposed **to** the air, which **dries** and hardens **the** cuticle, **and to** the floating particles **of** dust, **which** settling **into the** pores discolor and **irritate it, a** more careful **attention** will be required **to keep it smooth and in health than** that on any **other** portion **of the** body. **Cold water,** except in warm weather, should never **be used to bathe** the face, neck, and hands, **as it hardens** the skin **and** causes it to wrinkle. **Use only tepid** water and be sure **that it** is soft. Snow water is preferable when it **can be** obtained. **Bathe the** face, neck, **and hands the last** thing before retiring with **soft water and white** castile soap, soaking and gently bathing **it for several minutes.** Rub and press **the muscles of the face to make** them flexible, rinse off with diluted **alcohol or bay rum, then** anoint with a pomade, composed **of equal** parts of vaseline and almond **oil,** perfumed **with rose.** In anointing, rub the eyes inward toward the nose (at all **times do** this), **rub the** cheeks upward toward the crown of the head, to what Babbitt calls "**the** region **of har-**

dihood." This will prevent the deep downward lines forming at the corners of the eyes and mouth, and will also strengthen the eyes. Madame Recamié was in the habit of closing her eyes, four or five minutes, several times a day, claiming that by thus relaxing the muscles, it prevented the gathering of those wrinkles known as "crow's feet." The practice of shutting away all objects from the sight also rests the optic nerve. This advice is especially addressed to ladies approaching middle age.

The reason why some faces show more wrinkles than others, is partly because of the dryness of the skin. Anointing and rubbing will to a degree obviate this difficulty; but one important fact must be borne in mind in relation to this matter, and that is, the face is the index to the inner life. The soul writes its headings there. Every wrinkle on the face is a denotement of some strong emotion or passion; each having a different locality. Passing from every portion of the brain to the various muscles in the face are nerves called by physiognomists, poles, so that the soul expresses its passions and emotions upon the muscular cords as a musician would upon a musical instrument. Those which are used most constantly are the ones which will imprint the deepest furrows in the face. If the bad passions predominate then the lines of the face will turn downward; if the higher faculties are exercised they will curve upward.

Mirth draws the muscles of the mouth outward and upward, sexual passion outward and downward. Scorn and unkindly feelings draw the lip up at the corners of the nose in a contemptuous sneer, which in time becomes a fixed expression. Combativeness locates its poles on the forehead between the eyes. In anger these are contracted, giving rise to the frown of displeasure so familiar to all. Indulging in fretfulness, complaining, and faultfinding, gives an habitual elevation to the eyebrows, and long, deep wrinkles across the length of the forehead. In all cases the more strongly marked the characteristics in the individual the more decided the lines. It is said of Madame Patterson Bonaparte that she neither permitted herself to laugh or cry, knowing that both of these passional outgrowths would bring wrinkles, and wrinkles, to a degree, are blemishes. Women should learn not only to control the disposition that brings the wrinkles, but also to command the facial muscles, and under all circumstances prevent their telling so many tales. "He that ruleth his spirit doeth better than he who taketh a city."

This manifestation of the passions upon the face renders them contagious **to a** degree **to** those about us, even when no words **have** passed. **How** quickly it fires our combativeness when **we** come into the presence of an individual whose face is disfigured with a scowl of **anger,** although it may **in no** way be connected with us.

Macchiavelli was, in his study **of** human nature, in **the** habit of assuming **just** the attitude and expression of those whom he studied. **By** so doing, he **was** enabled to **fall** into the same train **of** thought, which gave him **immense** advantage over other men with whom he came in contact.

It is the duty of every woman **to** make herself as attractive **as possible,** and thus **be enabled** to **do a** greater **amount of** good in the world. The power **of Beauty is absolute. It is** worshipped **by** all, from the **least to** the greatest. Madame Recamié once said, when complimented upon her personal appearance by **a gallant admirer:** "You are **too** kind, sir, but **I** know **that I am fading** for **the** children no longer follow me **upon** the **streets** and call me 'Beautiful lady.' "

A woman possessing **a** high type **of** physical beauty, united with intellectual power, can, **if** she so desires **it,** have the **world** at **her** feet. Every woman has been **endowed by** Beauty **with one or more of** her attributes, which should be carefully preserved and enhanced, the same as any other endowment. It is just as reprehensible in women to allow themselves to contract slovenly slipshod habits in dress, as it is to fall into the use of common vulgar language. One loses his or her self-respect in proportion as they cease to regard personal appearance. The married woman who desires a continuance of conjugal felicity cannot afford to disregard in herself personal attractions. Men will admire attractive women wherever they find them. It is therefore the duty of women to employ every laudable means not only to enhance their beauty, but to improve the mind, that the grace of manner and appearance leaves nothing to be desired on the part of the husbands.

It may appear to the superficial observer that true love disregards all outward accessories and coquetries of dress. But this is not true. Love grows like what it feeds upon, and earthly love calls for a good share of earthly embellishments. When Want, Poverty and Disorder, clothed in rags and dirt, walk into the door, Love flies out at the window. Love is æsthetic and epicurean in his tastes and is easily disgusted. Husbands occasionally, when they get on an economical strain, strive to convince their wives that dress and genteel appearance have no sort of attraction for them. But don't you believe it! If you do not make yourselves sweet and pretty for them to admire, they will admire some other sweet and admirable woman; for it is the nature of men to love and admire something or somebody, and that something or somebody must be such as they can take pride in, be it woman, horses, or an establishment.

Woman embodies the ideal of Beauty, and man is a beauty worshiper. If wives would take one-half of the care to hold their husbands that they did to win them, few men would wander far from their allegiance. That which was attractive in the sweetheart will be equally so in the wife. There is usually a terribly letting down after marriage in all directions on both sides, not only going slipshod in dress but in temper also. What a wonderful revelation takes place in most cases. All restraint is thrown off, and often those traits, acts, and attentions which most charmed as lovers, are discarded as being no longer necessary. A husband should continue to be a lover all the days of wedded life, and the wife a charming mistress. Marriage is the beginning of courtship. The Persians have an adage worthy to be inscribed over the entrance to every home: "Familiarity discovers imperfections which reservedness conceals;" and as all men and women possess imperfections which are never rendered less odious by being brought to light, it would appear to be the duty of both husband and wife to conceal them, with a view to correcting them in time. The concealment which seemed necessary during courtship, is even more so after marriage, as crosses and cares accumulate. If deception was *at all* admissible before it certainly would be afterward. This care and consideration for personal appearance is not a one-sided affair; it is just as incumbent upon husbands to make themselves wholesome, and handsome, if it is in their power to do so; but as we cannot reach husbands, we give this advice where we give our love and sympathy—to the wives. We desire that women should cultivate and control themselves, and through a developed, refined, and well preserved organization, to control man.

Gœthe understood this principle, when he gave utterance to his eulogy upon women in the simple, eloquent words, "The female genius will draw us on high," and no man ever read human nature better, or loved women more devotedly than he.

But woman must first control herself, before she can control others. She must learn to set the heel of Will on the head of Passion, and thus hold it in abeyance.

MAL-POSITION.

The figure to be perfect, should be nearly flat across the shoulders. A sharp projection of the shoulder blades amounts to a deformity in any case, and what makes the sight more trying is that it has been induced, with rare exceptions, by the careless, slovenly habits of the individual in sitting, lying, and standing. This mis-shapen condition of the body may be corrected, even when one is quite advanced in years, by the practice of resting upon the stomach as long as possible in a perfectly horizontal position, dispensing with pillows; also the habit of sitting erect a few minutes, several times a day, on a hard-bottomed chair, with the arms thrown over the back. This, alternated by arm and breathing exercises, will soon cause the distended muscles on the back to contract, and thus draw the bones back in their place. Shoulder braces should never be used when the will can be called into requisition; but when the individual is destitute of that most important faculty, then of two evils choose the least, and put braces of brass, iron, or steel on them. Put them in the stocks! In the name of all that is lovely! do something to prevent the disgusting and distressing distortions which are now confronting us on all sides.

A "cross" or "squint" in the eye is considered a serious physical blemish, and one that is always corrected, if within the possibilities, but to the lover of symmetry it is absolutely refreshing and lovely, in comparison to curved spines and stooping shoulders; which with rare exceptions are unmistakable signs of weakness of the will, an indication of a lack of "vim," of mental and spiritual backbone, so to speak. School girls are likely to contract the habit of sitting in a stooped posture at their desks, and thus a mal-position is easily assumed, and as the bones harden it becomes permanent. Great care should be exercised over the young of both sexes in this respect. Nothing detracts so much from the bearing of a gentleman as "round shoulders" as they are called. An erect carriage is a patent of nobility, and impresses on first appearance as nothing else will.

But first and foremost, the health should be considered. A mis-shapen body is always an indication of depressed vitality, which is not only a serious drawback to the individual in a business point of view, but renders the system more open to diseases of all kinds, and less able to combat them.

CHAPTER XIV.

THE TOILETTE.

THE aphorism, "No excellence without great labor," is equally true in all departments of life, and unmistakably so in that of the toilette. When the skin on the face becomes subject to what is known as black heads, or pimples, a fine firm sponge should be used for washing, using diluted alcohol as a wash three or four times a day; never on any account using cold water to bathe the face, or expose it to sharp or cold winds. Gently press the skin with the balls of the fingers, but never squeeze or pinch the inflamed papillæ. Just before retiring each night take a wash bowl of warm soft water and bathe and soak the face in a gentle manner several minutes, then dry carefully with a soft towel and anoint with camphorated ice. In the morning moisten the sponge with diluted alcohol, and bathe the face, after which dust over the surface a little powdered starch or "Meen Phun." A pimply face requires the most tender treatment. An almost unfailing remedy for all blemishes of the complexion, pimples, moth, freckles, tan, and black heads, and one which the writer can conscientiously recommend, as being as harmless as honey or cream, is "Gourard's Oriental Cream," a preparation put up in New York City, but which almost any druggist would procure for their lady patrons. This preparation should not be classed with cosmetics generally, as it does not simply beautify, but restores the skin to a healthy condition. After thoroughly bathing the face with soap and water apply this "Cream" with a fine surgeon's sponge, allowing it to remain on.

The general health must be regarded, habits of constipation overcome if they exist, less salt used in the diet, and a freer use of esculent vegetables and acid fruits. Turkish or vapor baths should be used frequently. Bathe the body more and the face less; create an action in some of the multitudes of closed pores on the body, and give those on the face a rest. To keep the skin on the face fair, soft, and smooth, one must protect it from the sun and wind. The celebrated beauties of all times have not only carefully guarded their complexions, but have made use of all the accessories of the toilet in order to increase the smoothness and fairness of the face, neck, and hands.

A too free use of salt in the diet must be avoided by all persons taking little muscular exercise and having a defective action of the skin. The pernicious effect of an excessive use of this condiment is manifested in form of scurvy among sailors and soldiers who have been accustomed to the exclusive use of salt meats. If persevered in for a great length of time a disease makes its appearance, resembling leprosy somewhat, and after a time death ensues.

At times the limy principle of the salt is absorbed by the lymphatic glands, clogging and hardening them and destroying their usefulness, and where there is a constitutional tendency toward cancer, scrofula, erysipelas, or salt rheum, an excess of salt in the secretions aggravates, if not directly developes these diseases.

The skin is quite as active at night as during the day, therefore the night clothing and bedding should be hung or spread out in such a manner, that the fresh air may penetrate the meshes of the cloth and dry out the moisture exhaled from the pores during the night. The day clothing should be aired during the night, hanging them in winter near radiators, stoves or fireplaces, as heat is an effectual deodorizer; and when bathing is not frequently indulged in, the body-clothing should be changed two or three times a week, and bed linen fully renewed every week. Woolen blankets are preferable to cotton comfortables as they can be readily cleansed and aired.

THE SCALP.

For health, the scalp requires washing once a fort-night, with soft, warm water, castile soap, ammonia, or a fresh laid egg. Thorough brushing with a tolerably stiff brush is a daily necessity when the health and beauty of the hair is to be considered. The scalp is a mass of sebaceous glands, and cells containing the bulbous roots of the hair. It needs frequent rolling and pressing with the balls of the fingers to keep it flexible, and also, to press out the oil from the glands, and keep the hair soft and glossy, the extreme tips of which should be clipped once a month, only removing the split ends.

The latter treatment should be given the eye lashes two or three times a year, in order to make them even, thick, and long. Eyelashes are not given for their beauty alone; they help to shelter the eye from the too strong rays of light. The thicker they are, the better the protection afforded the eye; therefore, they should be, like the hair, cultivated, as they become broken and irregular in length like the neglected hair. To stop the hair falling out, saturate the scalp with strong sage or green tea and have the head frequently shampooed and vigorously brushed. Do not be alarmed if the hair does fall; it is by this means that it is renewed, the dead hair falling, and new coming in to take its place.

This occurs once every year just as it does in the case of animals in changing their coat, or birds moulting. Only give the scalp plenty of friction, and the new growth of hair will be forthcoming.

THE TEETH.

The mouth, throat, and teeth should be cleansed after every meal with clear water, and once a day the teeth brushed with white castile soap. Two or three times during the week a dentrifice, containing a slightly gritty substance, should be used to remove the tartar which ordinary cleansing will not affect. An excellent preparation for this purpose is equal parts powdered orris root, blood root, Peruvian bark, pulverized pumice stone, and prepared chalk. The brush must not be too stiff, and the water used should be tepid in all cases.

A small thin strip of pine wood should always be kept upon the toilette table, for the purpose of cleaning between the teeth and removing any particles of food which may have caught and remained after the meals. It will also help to remove the tartar at their base. Microscopic parasites are constantly forming on the teeth, which are only destroyed by alkali of some sort, soap being most effectual, Dr. Allen finding an absence of the tiny creatures in the mouths only of those persons who were in the habit of cleaning the teeth and mouth thoroughly after each meal, and using soap once in twenty-four hours. Nature never designed that the human family should lose their teeth any more than that animals should. They were intended to last a lifetime as much as the vertebra and skull; and when the proper attention is given the diet and the cleaning of the teeth in young children, toothache and decayed teeth will be as rare in the human family as they are in the animal species. If hot breads were dispensed with, and in the place of the fine flours more of the coarser breadstuffs were used, and the bread eaten only after it has become cold and thoroughly ripened, not only would dyspepsia, one cause of decaying teeth, disappear, but the enamel would be of a better quality and the teeth themselves firmer in their structure.

The teeth of children should be watched with assiduous care; the milk teeth removed as soon as loosened, and the permanent set overlooked from week to week for any symptoms of decay; which should on appearing be at once filled by a skillful and conscientious dentist. When, through neglect, large cavities have formed, placing the teeth beyond the skill of the dentist to redeem by filling, they should be removed, as decayed teeth are always unwholesome, not only to the individual possessing them, but to those persons coming in close contact, in such a manner as to take the fetid and diseased breath. **It** would prevent to a degree we fancy, promiscuous kissing, **if a** small portion of the collected matter on decayed and filthy teeth were put under the test of the microscope. The possessors of such teeth rarely escape some one of the forms of indigestion. **It** is neither safe nor pleasant to kiss the mouths of persons who have habitually **neglected the care** of their teeth, however sweet and agreeable they **may be** in every other respect. Few things **are** more disgusting in another **than foul** breath, which is largely due to this neglect; but **when it** results from catarrhal difficulty or an acrid state of **the** mucous coat of the mouth and throat, a gargle of **tepid water,** containing a few **drops of** carbolic acid, should **be used twice** daily, swallowing **a few drops** after **gargling;** but when offensive breath arises from catarrh alone, **the** lotion must be drawn up the nostrils until it can be tasted in the throat, using borax water frequently to cleanse the teeth and mouth. Another simple but **most** valuable remedy for bad breath, arising from an unhealthy **state of the** mucous membrane of the mouth, throat, and stomach, is lime water. Dissolve a bit of unslacked lime the size **of a hen's egg, in two** quarts of boiling water. After **it has** settled **drain it** off, and take **a** swallow three or **four times a day** just before meals—it may be taken in **milk, or if too** strong dilute with cold water. A little patience and care will **correct in a short time** this most disagreeable physical imperfection—a foul breath.

CARE OF THE EYE.

For weakness of the muscles **of** the **eyes,** bathing in cold salt water daily will be **found an** excellent remedy, always closing **the eyes and** bathing **toward** the nose. Continue bathing **each time until the muscles ache with** cold, after which **throw a cloth over the face and close** the eyes for **two or** three minutes, but **in case of** weakness or irritation of the optic **nerve,** regular intervals **of** rest is an imperative necessity—not **merely a** suspension **of** work for the **eye,** but **a** closing **of them** to shut **the** light and all objects **out,** for everything upon which **the eye rests** produces in that organ a vibration **of the** nerve filaments composing the retina—which **is an** expansion of the optic **nerve—therefore, rest can** only come by closing down the lid.

It was the practice of **Diana of Poitiers to** close her eyes long enough to count one **hundred, many** times during each day, as it kept them **dark** and bright, preventing the strong light from fading them. What **is** known **as** "nervous headache" is frequently **the** result of excessive use of the eyes. Almost everyone is familiar with the headache that comes from visiting picture galleries, expositions, and from inspecting goods on shopping excursions. The excitement of the optic nerve is communicated to all of the surrounding portions of the brain.

Light eyes are stronger than dark ones, which are seriously embarrassed by strong artificial light. If reading or other work is to be done at night, the light should be so arranged that it may fall over the left shoulder and never strike the eye directly in front. Opaque shades should never be used, as the condensed rays of light on the object on which the eye is engaged is so much stronger than that in the room that the retina is constantly irritated, as it would be by having strong rays of light thrown directly on it. As little labor as possible should be given the eyes at night, as it is then more difficult to adjust the lenses.

Bathing the eyes several times each day with a strong decoction of green tea, is an excellent stimulant to that organ.

THE EAR.

It would be supposed that every one would understand the care of the ear, yet a few hints will not be amiss on the subject of its treatment. The ear is no less wonderful in its mechanism than the eye, and the auditory nerve becomes inflamed quite as readily by loud and continued sounds, as the optic nerve does by constant sight and strong colors, and in both cases the brain would suffer. It is well to occasionally shut out sounds as well as sight. A constant unremitting din in the ear, has produced insanity in many instances, and no doubt is a prolific source of headache in the cases of persons exposed to continual commotion and noise. Mothers with large families of small children, teachers in public schools, women in factories, suffer more or less with irritation of the brain and consequent headache.

There is a certain amount of care due the external ear. We have seen in a previous chapter that nature has stored in the vestibule of the ear a quantity of wax, which is composed of an albuminous substance and a delicate oil. It is spread along the side wall of this vestibule, and the oil absorbed a little at a time by the tympanum, for the purpose, as we have seen, of keeping it soft and clear. If this wax is removed in any considerable quantity the membrane is deprived of the oil and to a degree becomes dry and loses its sensitiveness. It is a most injurious habit, and is often the aggravating cause of deafness. Nothing smaller than the point of the little finger with a wash cloth over it should enter the ear; tweezers, hair-pins, and other sharp, hard substances should never be used in cleaning this sensitive organ. As fast as the oil is consumed by the surrounding parts, the albuminous refuse is thrown off in form of white scales which can be removed by the finger with a clean soft cloth over it; and when the hearing is impaired from a hardened accumulation of wax, glycerine or some soft oil should be daily dropped in until the mass softens; then by pressing cotton wool into the aperture and leaving it for a time, then withdrawing it carefully, a portion of the wax may be removed, without taking it all, and leaving the membrane deprived of its lubricator. The ear can no more bear harsh treatment than the eye; only in extreme cases should syringing be resorted to. Temporary deafness is often induced by frequent douching; and earache in children may frequently be traced to the practice of removing the ear-wax constantly. Especially is this the case during cold weather. By this unwholesome treatment the ear becomes an incorrect medium of communication to the brain, and we have distorted hearing just as we have color blindness.

Not five out of every hundred persons after listening to a lecture would give anywhere near the same synopsis of the subject. No two repeat the same story exactly alike. The distortion of real facts, so distressing to the lover of truth, is oftener the result of imperfect hearing than a vicious disposition. It is wisest to

always distrust the ear a little, just as we are compelled
to do in the case of the eye, and not always accept as
truth what reaches the inner consciousness through
that much abused organ. Voltaire says, "Believe
nothing that you hear, and only half of what you see."
What a vast deal of misery we would all escape if we
lived up this axiom.

CHAPTER XV.

DISEASES AND THEIR REMEDIAL AGENTS.

HOME TREATMENT FOR COLDS AND SORE THROAT.

EVERY home should possess an apparatus for a vapor bath, the cost of which is a mere trifle. It consists of an alcohol lamp or "Pocket Cooking Stove," a common cane seat chair, and an oil cloth covering, with which to envelop the patient while taking the bath. Fill the lamp with alcohol, ignite it, and place over it a small vessel of water, which will boil in a few seconds. Place this heating apparatus on the floor, put the chair over it, fold two or three towels over the bottom of the chair; let the patient don a nightdress, over which throw a blanket, and take a seat upon the chair. Then let the attendant draw over the patient, chair and all, the oil cloth robe, which should be made with a hood and a draw string to draw it down closely about the face, in order to steam the head and neck in cases of sore throat, cold in the head, croup and congestion of the lungs.

The time required to produce profuse perspiration, differs in individuals. Those who perspire readily will

not require more than fifteen minutes, another may
need twenty-five or thirty minutes, but when the perspiration rolls off the face then the patient is ready
to come out. During the bath, administer hot drinks
freely unless there be a tendency to nausea and faint-
ness, in which case the drink should be cold. Iced
lemonade is excellent under such circumstances. The
room should be warm in which these baths are given,
and the patient bathed and dried—a limb at a time.
The moist body should not be exposed to the air. If
there is serious local inflammation do not bathe the
body, but slip the hand under the clothing and rub the
patient dry before removing the covering, then as soon
as possible put on a clean, dry, well warmed night dress,
and get the patient into bed and between woolen blankets if possible. One should always go to bed after a
bath of this sort, it matters not at what time of the
day they are taken.

One or two of these baths will be sufficient to break
up the most severe cold, and thus prevent a lingering
cough or protracted illness. The work must be most
thoroughly done, that is, the heat must be intense
enough to produce a copious perspiration, and also heat
the secretions thoroughly so that the patient will not
cool off too soon after coming out.

Another simple appliance, which every house should

have, is a couple of clean common bricks, which can always be kept hot in the range oven ready for an emergency. For cold in the head, sore throat, diphtheria, croup, hoarseness, etc., take a cloth wet in good strong vinegar, wrap it around the hot brick, and over that, one wet in water. Lay the brick on a small, thin board and let the patient hold it close to the mouth, inhaling the steam, covering the head with a blanket to prevent the steam escaping. Continue to inhale as long as possible, then raise the cover and take in a supply of fresh air, then return again to the steaming, continuing this treatment until the congestion has been subdued. It may require several trials, but it is a sure cure, if taken in time, for most of the difficulties above mentioned. A person can be treated in this manner while in bed or while sitting up.

In diphtheria the patient will require stimulating, for an adult a table-spoonful of best brandy or whiskey must be administered each hour, until the fever subsides and the pulse becomes stronger and more regular. The spirits may be diluted and prepared as a drink and given in small quantities at stated intervals. Strong lemonade is also most grateful to the feverish palate as well as cooling to the fever. The best mode of preparing it, is to pour a quart of boiling water over an ounce of gum arabic. After it has dissolved and cooled, drain off carefully, and use this water to make the lemonade, as the gum arabic nourishes the patient and furnishes a needed element to the inflamed and overtaxed membranes.

A free use of the "flour of brimstone," common sulphur, is also highly recommended by many of our best practitioners, both in the cases of diphtheria and membraneous croup. Sprinkle a small quantity of the dry powder down the throat and on the tongue, the patient mixing it with the saliva and slowly swallowing it.

In the cases of children suffering from croup when they will not bear the steaming from the bricks, apply to the throat and chest compresses wrung from salt water, just as hot as the flesh will bear them, covering all with hot dry flannels, having previously anointed the surface with vaseline, or oil of some sort. The compresses should be frequently changed in order to keep up the heat. Put the feet and lower limbs in hot water until the flesh appears red. One-fourth of a teaspoonful of pure vaseline should be administered once every ten or fifteen minutes until the breathing becomes easier. Equal parts of powdered sugar and alum sprinkled down the throat is an almost unfailing remedy. This should be applied as often as once in ten minutes until relief is obtained.

Children with **croupy** tendencies should be carefully watched **and not** permitted **to** go with cold **feet,** but should **during the fall,** winter, and spring be clothed in flannel, **the feet** bathed daily in **hot salt** water and vigorously rubbed. There should also **be a** frequent change of stockings.

When these precautions **are** observed, croup will disappear, for it **is in nine** cases out **of** ten **the result** of exposure, the child going **for days with** cold **feet** and a chilled surface. All croupy persons are greatly benefited **by the free use of** honey **as** an article of diet, **and when it cannot be** used, syrup should take its place. **A certain amount of saccharine in the diet is** just as **essential** as bread, especially in **the case of the** young. **It is heat-producing** and assists in **warding** off colds. **It is** a legitimate appetite, in children, that demands sweet, which should be reasonably indulged, not supplied in highly **colored candies,** but **in** a generous **allowance of syrup,** honey, and sugar with the food, particularly if **there are any** indications **of** colds or croup.

CONTAGIOUS AND ERUPTIVE FEVERS.

The following **is an** extract from **a** noted English **publication:** "Contagion, then, consists physically of **minute solid** particles." (Disease **germs.**) " The pro-

cess of contagion is the passage of these from the
bodies of the sick into the surrounding atmosphere, and
in the inhalation of one or more of them by those in the
immediate neighborhood. If contagion were a gaseous
or vapory emanation, it would be equally diffused
through the sick-room, and all who entered it would, if
susceptible, suffer alike and inevitably. But such is
not the case; for many people are exposed for weeks
and months without suffering. Of two persons situated
exactly in the same circumstances and exposed in
exactly the same degree to a given contagion, one may
suffer and the other escape. The explanation of this
is, that the little particles of contagion are irregularly
scattered about in the atmosphere, so that the inhalation of one or more of them is purely a matter of
chance, such chance bearing a direct relation to the
number of particles which exist in a given cubic space.
Suppose that a hundred germs are floating about in a
room containing two thousand cubic feet of air. There
is one germ for every twenty cubic feet. Naturally the
germs will be most numerous in the immediate neighborhood of their source, the body of the sufferer; but
excepting this one place, they may be pretty equally
distributed through the room, or they may be very
unequally distributed. A draught across the bed may
carry them now to one side, now to the other. The

41

mass of them may be near the ceiling, or near the floor. In a given twenty cubic feet, there may be a dozen germs, or there may be none at all. One who enters a room may inhale a germ before he has been in it ten minutes; or he may remain there for an hour without doing so. Double the number of germs and you double the danger. Diminish the size of the room by one-half, and you do the same. Keep the windows shut, and you keep the germs in; open them, and they pass out with the changing air. Hence the importance of free ventilation; and hence one reason why fever should be treated, if possible, in large, airy rooms. Not only is free ventilation good for the sufferer, but it diminishes the risk to the attendants.

"We see in this, too, the reason for banishing bed-curtains, carpets, and all unnecessary furniture from the sick-room in cases of contagious fever. The germs are apt to adhere to such articles, and so make them the means of conveying the disease to others.

"All organisms consume in their growth *nitrogen* and *water*. Those with which we are now dealing are no exception to the rule. Growing in the system, they must get these elements there. But nitrogen and water are the chief materials required for the nutrition and repair of the various organs and tissues of the body. The propagation in it of millions of organisms, having wants identical in the main with those of its own tissues, must cause serious disturbance. And so it does. This disturbance declares itself by that aggregate of phenomena to which we apply the term fever.

"An organism which thus grows in and at the expense of another is a *parasite*. One of the peculiarities of parasites is that they flourish, not in any part of their host, but only in some *particular* organ or tissue, which is called the *nidus* or nest of the parasite. The organisms with which we are now dealing (the poisons of the eruptive fevers) show similar peculiarities. Each has its own nidus, its own localized habitat, in which it is propagated, and out of which it ceases to be reproduced. The poison of small-pox has its nidus in the deep layer of the skin; hence its characteristic eruption. That of scarlet fever, in the superficial layer of the skin and in the throat; hence the rash and the sore-throat of that disease. That of measles, in the skin and in the mucous membrane of the air-passages; hence its characteristic symptoms. That of typhoid fever, in the glands of the intestine; hence that disease consists of fever and of ulceration of the bowel.

"The contagiousness of a given eruptive fever must be directly as the number of germs which, in a given time, pass from the body of a sufferer into the surrounding atmosphere. This, in its turn, must depend on the seat of the propagation of the poison, and on the relation which this bears to that atmosphere. In small-pox, scarlet fever, typhus fever, and measles, the seat of this propagation is the skin and mucous membrane of the air passages; it is, therefore, in direct, free, and constant communication with the external air. The poisons of these diseases are accordingly freely given off into the atmosphere of the room in which the sufferer is, and they themselves are highly contagious.

"In typhoid fever, the poison is propagated in the bowel, and is thrown off with the discharges from it. It thus passes from the system in a manner and in a combination which insure its speedy removal from the neighborhood of the sufferer. The typhoid-germs are there, but they are mingled with discharges which may be removed, and as a matter of course are removed, before the germs can pass off from them into the surrounding atmosphere. The seat of the propagation of the typhoid-poison has no direct relation with this atmosphere; germs cannot pass directly from the one to the other; the disease, therefore, does not display the property of contagiousness.

"The danger in typhoid fever is not contact with the person of the sufferer, but contact with his stools. If these are properly managed and disposed of, the disease can scarcely spread. But, if they are allowed to pass into drains which are imperfectly trapped, inadequately ventilated, or insufficiently flushed, or if they are carelessly thrown on the ground, or allowed to percolate through the soil into drinking water, then one case of typhoid fever may give rise to many others. But the communication of the disease is not direct, by contact; it is indirect—by infection of drinking water, or of an atmosphere which may be remote from the person who is the source of the poison. A case of typhoid fever is introduced into a locality. The stools are thrown out on the ground, or into a cesspool, where they percolate through the soil into a well. The person who drinks water from that well runs a greater risk than one who sleeps in the same room as the sufferer and is in constant attendance on him.

"The practical outcome of all this is, first, that the mother may nurse her son, the wife her husband, the sister her brother, without the risk involved in the case of typhus or scarlet fever; and, second, that there is little or no danger to the other inmates of the house, if its sanitary arrangements are perfect and the stools properly managed.

"On this view of the nature and mode of action of contagion, it is easy to see, not only how the process of contagion and its varying phenomena may be explained, but how, by care, much may be done both to prevent

the poison from passing into the atmosphere and to
diminish its chance of acting after it has got there.
We have only to consider what is the chief channel by
which contagion gets exit from the system, to know by
what means we are most likely to prevent its passing
into the surrounding atmosphere. In typhoid fever the
poison passes off in the stools; and what we have to do
is to see that these are promptly and properly disinfected and disposed of. In small-pox, scarlet fever,
typhus fever, and measles, it is eliminated by the skin,
and we cannot altogether prevent its getting into the
atmosphere; but, by frequent sponging with some dis-
infecting fluid, or even with plain water, many germs
may be arrested in their outward course.

"The apostolic mode of anointing with oil is also an
efficacious way of fixing and arresting the germs: it is
especially useful during convalescence from scarlet
fever in fixing the particles of peeling skin, which are a
source of much danger. They are dangerous because
they contain the germs which have been produced in
them. What we see happen in the larger particles of
the skin happens also in many of the much smaller
particles of contagion.

"By the adoption of these various measures, and by
having the room well ventilated, much, very much may
be done to check the spread of contagious fevers. The

matter of which organisms are composed is one of the most perishable things in nature. Contagion is no exception to the rule. By exposure to the air much of it is destroyed; hence such exposure is one of the best of all disinfectants.

"Sanitary science has done much to show us how some of the diseases with which we are now dealing might be extinguished, and how all of them might have their prevalence greatly diminished. It rests with those who have such ailments in their houses to carry into effect the measures calculated to destroy and get rid of the poison, before it has had time or opportunity to be a source of danger to those around. But the adoption of proper measures presupposes a knowledge of the nature of the poison with which we have to deal, and of the manner in which it passes off from the system. In not one is this knowledge more necessary than in typhoid fever; in not one are the measures which such knowledge dictates more easily applied or more likely to be effective. But, to regard typhoid fever as contagious in the sense that small-pox and typhus fever are so, is to divert attention from the true source of danger, to lead to the adoption of measures which are uncalled for, to the neglect of those which are urgently required; is to cause unnecessary concern to the sufferer and his friends, and to deprive him and them of the mutual comfort and solace which a little daily intercourse affords. The peculiarities of the illness may be such as to make it right to exclude the friends; but isolation is not requisite for the same reason that it is so in typhus.

"One more point. The receiver as well as the giver of the poison has something to do with the determination of its action. Not every person into whose system a germ passes, necessarily suffers from its action. A man who has had small-pox, for instance, is no longer susceptible to the action of its poison, and why? Not because the poison can not get into his system, for we can make sure of that by inoculating him with it, but because during the first attack, the nidus, the special material necessary to its propagation, was exhausted, and has not been reproduced. This immunity from a second attack is a general characteristic of the eruptive fevers; individual exceptions there are, but the rule is that one attack confers immunity from a second.

"A germ does not act unless it *reaches its nidus;* it may enter the system, make the round of the circulation, and again pass out without ever coming in contact with its nidus, and therefore without doing harm.

"The more widely the nidus is diffused the less likely is this to happen. In small-pox, in scarlet fever, and in measles, the nidus is widely scattered. In none of them is a germ likely to make the round of circulation more than two or three times, without being conveyed to its nidus.

"In typhoid fever the nidus is situated in a limited portion of the bowel, the sole route to which, by way of circulation, is through an artery the size of a crow-quill; a typhoid-germ may be taken in through the lungs, and may make the round of circulation two or three dozen times without being likely to enter that particular vessel. The more often this may occur the greater the chance of its being thrown off from the system without acting. But, if the typhoid germ be taken in through the digestive organs, it is brought into direct contact with the seat of its nidus, and can scarcely fail to act. Hence the great danger of drinking water or milk contaminated with the typhoid poison.

"The glands which constitute this nidus are not equally prominent and active all through life. In infancy they are quite rudimentary. At two or three they begin to grow, and gradually increase in size, and presumably in functional activity, till the age of puberty. They continue to be very distinct for twenty or twenty-five years. After forty they begin to get less, and gradually diminish, till at seventy they have dwindled away so much that they can no longer exercise any active function. Their period of prominence and

of functional activity corresponds exactly to the period
of susceptibility to the action of the poison of the
typhoid fever. That disease is extremely rare in
infancy; from two to six, or seven, it is more common,
but is generally very mild. At fifteen or sixteen commences the period of greatest liability to it; and from
that age until thirty-five and forty it is very common
and very fatal. After forty-five it begins to decline in
frequency and severity, and goes on declining as years
advance, till at seventy the liability to it may be
regarded as practically worn out. When it occurs in
advanced life it is generally mild, but its occurrence
then is as rare as in infancy. Increased and diminished
susceptibility to the action of the poison of typhoid
fever correspond exactly to the increase and diminution
in the size and functional activity of the glands which
constitute its nidus.

"Regarding the typhoid-poison as a parasite whose
nidus is in the glands of the bowel, we are led to the
conclusion that the disease to which it gives rise,
though undoubtedly infectious, can scarcely be contagious. We know from our experience that it is not so;
for it never spreads in hospitals, and attendants on the
sick suffer no more than other people.

"The difficulty has been to reconcile these facts
with the reproduction of the poison in the system.

The source of this difficulty is the **rooted belief that** this reproduction takes place in **the blood.** On this view **all** the eruptive fevers ought to be equally conta**gious.** But let us once **adopt** the view that the poisons of the eruptive fevers **are** *parasites*, **and** that the seat **of** the local lesion of **each is** the nidus **of its** parasite, and therefore **the seat** of its propagation, and the whole difficulty vanishes. We at once **see why each** has a definite period of duration, why **one attack** protects against a second, why each has **its own char**acteristic lesion, why each presents such **varying** degrees **of** severity, and why they possess different degrees of contagiousness."

The importance of frequently and thoroughly **disin**fecting all human habitations should be **more** fully understood by all, especially where contagious diseases exist. There should **also be a** better knowledge of **the** manner in which **the** various contagions are conveyed, in order to guard **against** the spreading of such diseases. One disease is communicated by the breath, another by the bodily exhalations, another **by** the passages, **as** in the cases of typhoid fever, so that the danger of infection varies accordingly. **In** case of typhoid, the passages should be removed far from all springs, **wells,** **and** water courses supplying the **needs of man and** **animals,** and deeply covered with lime, charcoal, or dry earth, all of which are first-class disinfectants, and within the reach of all.

In case of scarlet fever, bedding and clothing, after being removed from the patient, must be at once thrown in boiling soap-suds, and allowed to boil five or ten minutes.

Of course all clothing from small-pox patients should be burned.

A small vessel of boiling vinegar will not only be found to be a perfect deodorizer, but a powerful disinfectant, and should be resorted to in all cases where there are eruptive fevers. Saucers of bromo chloralum should be placed in various positions near the sick bed. The air should be continually renewed and the foul air forced out, therefore an open fireplace in a sick room is indispensable, but where this is not practicable the stove door, if there is a stove in the room, should be left open constantly, as it serves the purpose of an open grate in cleansing the air. Vessels of water, frequently renewed, also serve the same purpose.

All disease germs, or the parasitic formations, have a strong affinity for water, therefore frequent baths and packs are of the first importance as remedial agents. Dr. Hallier states that he has always found large quantities of the parasites in the exhalations from the body after the use of these applications, and there can be no doubt but that the use of baths and packs assist the egress of these tiny creatures through the pores.

A judicious use of carbolic acid in both the baths, gargles, and enemas, is indispensable in the case of fevers.

Quantities of disinfectants for use may be thus stated: For privies or sewers, a pound of sulphate or chloride of iron or chloride of lime, diffused in a gallon of water, will answer for a very large amount of foul material. Burnett's liquid contains twenty-five grains of chloride of zinc in each fluid-drachm of water. A pint of this in a gallon of water will be strong enough for use. For water-closets or bed-pans, Labarraque's solution of chloride of soda, a fluid ounce in a quart of water; or Condy's liquid, ten grains to a quart of water; or carbolic acid, twenty grains to a pint. A seventy per cent of this last named substance is often used also. Drinking water is best purified by filtration through charcoal, but it may be improved, when containing an excess of organic matter, by a small amount of Condy's liquid, enough to make it very slightly pink in color in a strong light. Occupied rooms may be disinfected by fresh chloride of lime, placed about in saucers in convenient places to give off chlorine.

But the best known disinfectant, perhaps, is *heat*. Pliny says, "There is in fire itself a medicating power," so that the efficacy of heat was understood centuries ago. Dr. Henry, of England, performed a series of experiments by which he proved that the contagion from small-pox, typhus and scarlet fevers is destroyed by a temperature of from one hundred and forty to two hundred degrees Fahrenheit. This accounts for the marvellous benefits derived from the Turkish bath, by patients suffering from the various eruptive fevers. Intense heat and intense cold are both destructive to parasitic life. Heat is always within the reach of everyone, and can be applied to both houses and patients.

SCARLET FEVER.

The patient should be kept free from draughts of *cold* air. The room must be light and of sunny exposure if possible, thoroughly aired and disinfected, which in cold weather can only be accomplished by a high degree of artificial heat. A vessel of strong vinegar should be kept constantly simmering in the room, and a saucer of chloralum kept near the bed. The one important point is to keep the patient warm, while continually changing the air in the room. Administer a hot foot-bath on the first appearance of the disease, put the patient in bed and pack the body from the chin to the hips in sheets wrung from water as hot as can be borne, over which place hot, dry flannels, surrounding the patient with bottles of hot water to keep up the heat, allowing them to remain from twenty to thirty minutes. These packs should be renewed until the rash is fully established on the surface.

On no account allow the bedding to become moist, or the cool air to strike the body. As previously directed, the packing clothes must be at once removed from the sick room and thrown in boiling suds.

The entire surface of the body must be oiled with some soft oil or vaseline; the hand of the nurse being passed under the clothing while anointing. Allow the patient a free use of iced gum arabic water, alternating with bits of broken ice. No other nourishment than good, pure milk should be administered until the fever has abated. Great care should be observed to prevent the convalescing patient from taking cold. A tepid salt water sponge bath should be given once a day, keeping up the anointing. Clothe the patient in flannel, varying the thickness to suit the season. The mouth and throat should be cleansed several times a day with borax water, or a solution of carbolic acid and water, and each time after so doing let the patient swallow a few drops.

HEADACHES.

In all cases of headache, regardless of the provoking cause, there is an overcharge of blood in the cranial vessels and a receding of that fluid from the extremities, giving rise to cold feet and hands. The first consideration in the treatment of this difficulty is to restore the circulation, then absolute rest is the next condition to be observed, just as we would guard inflamed eyes, a burn, or wound, to give the recuperative powers time to subdue the irritation in the one, and form new fibres and tissues in the other. The delicate membranes which enfold the brain become engorged with blood as the white membrane of the eye does in case of inflammation of that organ. The light should be shut away for a time by some light bandage, and a few magnetic passes made from the head downward by some strongly magnetic person, alternating these passes by holding the head between the hands, the right hand at the base of the brain, and the left over the front portion, allowing them to remain only a few moments, and frequently cooling the hand in cold water. Each time upon removing the hands pass them from the head off along the arms, toward the hands of the sufferer, taking care to keep the feet and hands of the patient warm.

Drinking copiously of hot water from time to time, is a simple and excellent aid in restoring the lost balance. Also counter irritants, such as extremely hot, or extremely cold water applied along the spine. Mr. Brown Sécard orders that the spine be rubbed with ice in extreme cases.

This treatment must be varied to meet the needs of the individual temperament. Hot applications may do for one, and only cold for another, but sharp friction that will bring the blood to the surface and keep it there will give relief in nearly all instances. Sometimes a mustard leaf applied low on the neck will relieve the suffering.

Another simple and efficient remedy is a small sack of hops wet in boiling vinegar and bound on the head, and when that is inaccessible, towels wrung from hot salt water, applied as hot as the part can bear, enveloping the entire surface with dry flannel, renewing the heat frequently.

These are merely assistants after the attack has come on. In order to bring about a cure the aggravating causes must be removed. It may result from indigestion, constipation, menstrual irregularity, undue excitement, excessive mental or physical labor, whichever one it may be, must be corrected, before the headache can be thoroughly cured. A daily, hot salt

43

water, foot and hip bath should be resorted to, rubbing and pinching the feet until the blood flows into the weakened vessels, and the feet become permanently warm.

As much capsicum as can be taken upon the point of a penknife, mixed with a little cream, and taken before each meal, will often perform a cure when the headache arises from indigestion.

Persons of sedentary habits who are subject to this distressing ailment, should form the practice of taking exercise in the open air daily, if only for a short time; also light gymnastics which would tend to call the blood away from the overcharged brain to the muscle. A cold salt-water hand bath, combined with thorough friction, with coarse towels and flesh brush, would be advisable each morning. This bath must be a mere dash, to serve as a tonic, and need not consume more than three minutes at the outside. It will restore a normal tone to the capillaries and nerves on the surface of the body. The room in which such a bath is taken should be some degrees warmer than the water used, but in all cases the *friction* must be sharp and the skin made to glow. We would advise those persons unaccustomed to this sort of treatment to bathe only the breast and arms at first, then gradually extend the area over the body. This, like all the treatment, must be persevered in, if the best results would be obtained.

If the headache comes from indigestion, the diet must be regarded; if constipation, then that will need to be corrected, as will the menstrual difficulties and nerve irritation, all of which have been treated upon in various chapters. There is one indispensable to good health, and that is *warm feet*. No one can possibly possess the former without having the latter, and in order to overcome the habit of cold feet, a daily foot treatment must be resorted to. The feet must be rubbed, pinched, masculated in such a manner as to bring the blood into them. One of the very best methods for stimulating the circulation is to rub the feet briskly over the carpet after taking a bath. The stockings must be changed daily, the shoes should be loose, thick, and warm. A given amount of exercise on foot is absolutely necessary.

RHEUMATISM AND ITS TREATMENT.

Rheumatic tendencies are most frequently inherited, being temperamental conditions which unfavorable surroundings develop into forms more or less aggravated. There are several causes which tend to greatly favor the development of this distressing malady. In malarious countries the most common is liver difficulty and its concomitant, dyspepsia, both of which must be removed before any improvement can be anticipated; in variable climates the difficulty is intensified by the sudden and extreme changes of temperature, which cause rapid alternation in the perspiratory system by first heating and then chilling the vascular structure of the skin, inducing imflammation of the internal membranes, especially the periosteum and synovial capsule, two delicate membranes, the former sheathing the bones, the latter lining the joints. This irritation gives rise to what is known as bone and joint rheumatism.

For this type the Turkish and vapor baths are of the first importance, if taken in time, as they assist the skin and kindeys in the performance of their work, and when those two channels are kept open and unobstructed, the rheumatic difficulty will, in most cases, disappear.

Persons suffering from this disease should wear flannel under-garments, which should be changed frequently. The better way would be to have two sets in use, changing every alternate day, hanging the discarded suit where it can be thoroughly aired meantime. The same treatment should be given the stockings, and the feet bathed daily with strong salt water. In most instances a reasonable use of lemons and cider vinegar will be found most bene-

ficial, especially where the disease results from bil-
iousness, for generally there is a superabundance of
alkali in the system, which acid neutralizes. It is
said that bee stings will cure rheumatism. The writer
has never tested this remedy and cannot, therefore,
vouch as to its efficacy, which is claimed to result
from the acid in the poison of the sting. But there is
a simple remedy which we have tested fully, and one
which we can conscientiously recommend to all persons suffering from indigestion, cold condition of the
system, and rheumatism. It is capsicum, the only
stimulant known which has no unpleasant reaction.
In extreme cases it may be administered three times a
day, just before meals. The dose may be slightly varied
to suit the condition of the patient, as some persons
are more readily influenced than others by medicines.
The usual dose for an adult would be what could be
taken upon the point of the large blade of a pen-knife.
Mix it with thick sweet cream, which relieves the
unpleasant biting and smarting in the throat. There
must be daily rubbing of the joints and parts affected
by warm, strong magnetic hands, and friction from a
flesh brush over the entire body, together with sun
baths whenever they can be obtained. A shaded,
damp, and sunless house is certain death to the rheumatic patient. Such conditions are just as sure and

destructive in their effects as that continual exposure to sunshine is certain death to rheumatism. The rheumatic sufferer should follow the sun, and spend **as** many hours in its direct beams **as possible.** A high temperature is most agreeable, and best suited to this class of persons.

FAINTING.

In cases of fainting lay the patient on her back, in **a** horizontal position, loosen the clothing and apply cold **water** to the face, chest **and** pit of the stomach, rubbing the hands **and** feet, and frequently applying **salts and** camphor to the nostrils. Persons predisposed **to** fainting should **take** breathing and **arm** exercises daily **in** the open air and study self-control, for this difficulty like hysteria is greatly under the control of the will. Crowded and ill ventilated rooms should be avoided by those disposed to this weakness.

HYSTERIA.

Hysteria, **in its** initial stage, is **a** disorder of the **emotions, and** influenced by the passional nature, and is wholly under the control of the will. Properly, it cannot be called a disease any more than laughing or crying. It is produced by uterine disturbances, either from starvation or over-exercise of that organ. In unmarried women from unallayed sexual excitement or sexual abuse, in married women excessive coitus or repeated attempts to prevent conception.

Being entirely a nervous difficulty, medical agents have little or no effect upon it. A permanent cure can only be established by removing the aggravating causes, developing the will, and cultivating the intellectual faculties. When from repeated paroxysms the nerve ganglia becomes greatly inflamed, leaving the patient liable to attacks on the slightest excitement, then outside agencies will be required to assist in restoring the lost balance. The hip bath, hot vaginal douches, Turkish baths, will be, except in rare cases, sufficient to establish a cure; but the will-power is superior to all other agents in difficulties of this character.

In the latter part of the last century, hysteria became so prevalent in the convents of certain portions of Italy, that some of them were entirely broken up. An expedient was resorted to which effectually cured the difficulty. A physician was consulted who prescribed stripping the patient and branding the body with a hot iron. To make it more impressive, the proclamation was made in the various convents and the iron exhibited to the nuns, after which, strange to say, not one case occurred.

CANCER.

Cancer generally locates itself in the glandular system, and aocording to M. Banphieting, the eminent French scientist, is caused by a microscopic parasite or animalcule which he claims is always, upon close examination, found in cancers; consequently, if his theory be correct, whatever will tend to destroy these minute creatures will hasten a cure.

The diet has an important influence upon this disease. All articles of an inflammable nature should be avoided, such as an excessive use of salt and highly seasoned food. Pork or other fat meats should be dispensed with entirely, and a liberal use made of acid fruits of every description, especially lemons and oranges, also drinking freely of the various mineral waters. The avenues of purification must be kept open, the pores of the kin, the bowels, liver and kidneys. This can be accomplished without the aid of drugs, by proper diet and muscular exercise, together with a thorough course of bathing, employing the Turkish, vapor, and mineral baths, which as curative agents in all cases of blood disease stand pre-eminent, as has been demonstrated by the mineral baths of Germany, especially in cases of scrofula, erysipelas, and salt rheum, which are all members of one family.

In the case of cancer, the knife should never be resorted to, as the shock to the nervous system is very great, and in the removal of the tumor only the effect and not the cause is dealt with, the poison still remaining.

Some wonderful results have been obtained through what is known as the Thermal cure. The treatment consists of throwing upon the cancer a steady jet of heat, which effectually destroys the life, after which the dead flesh sloughs off, allowing the healthy flesh to heal. There are also certain drugs which by being externally applied will destroy the abnormal growth and in many instances save the life of the individual.

But the only permanent cure would be in building up the system and keeping the blood in a healthy condition, not through the aid of drugs, but through those avenues designed by nature for that purpose. All persons having a hereditary predisposition toward blood diseases of the cancer family should make daily use of stewed cranberries, during the season of this fruit.

In females, the mammary gland is most subject to these growths, arising, doubtless, from bruises, compressions, congested milk glands, etc. This gland should never be permitted to hang by its own weight, as such a position produces strangulation of the blood-vessels, tending to induce morbid growths. There is an intimate connection between the ovaries and the breast. Inflammation in the former produces, many times, painful hardening in the **latter.** Particularly is this true toward the latter part **of** the change of life, often giving **rise to** apprehensions **of** cancers **and** tumors, and **by a** sufficient amount of **worrying over** and squeezing, **might** be coaxed into such, but if undisturbed will gradually disappear, **but where it** continues beyond a reasonable **time,** hot applications, as hot **as can be** borne, should **be** applied **to** stimulate the absorbents **and do away** with the **hardening** and inflammation. Fomentation performs **the** same mission **to** the breast that it **would** to inflammation in any other part of the body.

It **is** claimed by many that hot cranberry poultices will destroy **these** diseased formations.

More people die from imaginary cancers than from real ones. Women especially are apt to torture every little itching papillæ or inflamed lymphatic gland into one **of** these terrors.

SCROFULA, ERYSIPELAS AND **SALT** RHEUM.

The same general treatment advised for cancer would **also** apply to the above diseases. The only difference between them is, that cancer **attacks** the larger glands, while scrofula, erysipelas, or salt rheum, manifests itself on the surface, and is therefore **more readily** reached by external remedies, such as baths, anointing, emollients. Frequent hot baths, the Turkish bath always when it can be obtained, must **be** resorted to, and **a** thorough daily anointing of the entire surface.

In case of erysipelas, poultices of grated beet or stewed cranberries, applied to the inflamed parts, give almost immediate relief. **If** people were **in the** habit of bathing more frequently, and making **a freer use of** lemons and acid fruits generally, **erysipelas and kindred diseases** would cease to afflict mankind.

In case of salt rheum only **warm water** should be used for bathing the parts, using vaseline **as an** ointment several times **a day.** *Litlle or no salt, and no pork should be eaten.*

DYSENTERY AND DIARRHŒA.

In case of chronic diarrhœa, double flannel abdominal bandages must always be worn, and the bowels bathed daily in strong salt water. Dip the hand **in** the water, bathe and gently rub the parts until **the** surface glows. **Once a** day a tepid salt water enema should be used. Avoid **a too** free use of fluids. The breadstuffs should be dry, such as crackers, dry toasts, etc., taking little or no fluid with the meals, using capsicum freely on all **savory** articles. Avoid cold drinks, ice water, etc.

Form the habit of chewing (although it may be vulgar) some one of the native gums, pine, spruce, tamarac, swallowing the saliva. The resinous oil contained in these gums is most healing and stimulating to the weakened mucous linings.

A clergyman, formerly a chaplain in a Wisconsin regiment, informed the writer that he had over two hundred certificates from men who had been cured of chronic diarrhœa, simply by chewing these gums.

Diarrhœa is occasioned by a negative and lifeless condition of the mucous linings, therefore whatever would tend to stimulate those parts would bring relief. The diet should be largely composed of those articles less cold in their nature and more easily digested.*

Dysentery is the exact opposite of diarrhœa; it represents fever; the mucous linings are inflamed and at times lacerated, giving rise to the disease known as bloody dysentery. This difficulty requires soothing remedies; nothing of a stimulating character should be used. No food should be taken for a time, except milk to which has been added lime-water in proportion of a table-spoonful to a goblet of milk. After each evacuation an enema must be administered. In a pint of tepid water stir one table-spoonful of wheaten flour and inject into the rectum, encouraging the patient to retain

---

* See article on Digestion.

it if possible. Warm applications over the bowels give relief. The patient must be kept perfectly quiet and free from excitement.

CONVULSIONS IN CHILDREN.

The predisposing causes of convulsions in children are inherited, but the difficulty is developed mainly through indigestion. Their diet must be most carefully guarded and nothing of an indigestible nature allowed them. This class of children should have plenty of out-door exercise, and, as far as possible, kept out of school until they overcome the constitutional tendency.

When the attack comes on, the child should be given an emetic of tepid water and mustard, as relief is experienced after copious vomiting, and the body immersed in a bath of hot salt water. Then wrap the child in flannels, keeping up the friction until there is a glow over the body, always resorting to this treatment whenever the paroxysms come on.

Epilepsy is more difficult in its treatment, and is usually the result of long-continued abuse of the generative functions, such as Onanism and solitary abuse, and rarely, unless inherited, attacks young children. Where the practices inducing these diseases have been of long standing there is no cure, but if corrected in time the difficulty disappears.

LIVER AND SPLEEN DIFFICULTIES.

The office of the liver, as we have seen, is to secrete bile. The blood in its passage through this organ leaves this refuse in the innumerable minute glands arranged for this purpose. From these glands the bitter fluid is conveyed through tiny ducts to what is known as the gall cyst or bladder; from that receptacle it is emptied into the lower bowels, creating in their walls a spasmodic action, giving rise to a desire for an evacuation of the fæces. This is probably one of the most important offices of the bile, although we have seen in a previous chapter that in conjunction with the pancreatic juice it has a work to perform on the partially digested food. There is a steady accumulation of this fluid in the liver which must be as constantly expelled. The liver like all other glands has no contractile power of its own, but depends upon the contraction of the surrounding muscles to assist in expelling the secreted bile. The diaphragm, stomach, and abdominal muscles perform this work by a constant and gentle pressure. If the clothing is worn tightly in the region of the stomach and diaphragm, their motion is impeded, the liver becomes overloaded with bile and hardened, giving rise to what is known as inactivity of the liver. The best remedy for this condition would be rolling, kneading, and pressing to assist the engorged organ in freeing itself. A compress wet in strong vinegar and worn during the night over the hepatic region is also an excellent remedy. Both the mechanical treatment and the compresses should be worn until the action is established, as drugs rarely reach those parts.

SPLEEN AND ITS OFFICE.

The spleen is a soft, spongy organ of a deep violet red, situated in the left side below the diaphragm and above the colon, and in front of the kidneys. In health, it is four and a half inches long by two and a half wide, and its weight about eight ounces, Its coat is composed of various elastic tissues; its interior is a pulpy mass, containing multitudes of grayish, semi-transparent granulations. This organ is attached to the general circulation by a plexus of veins and capillaries which ramify the substance of the splenic pulp; the upper portion is called the head, the lower portion, the tail. The exact functions are not known, but from the best opinions we learn that it serves as a reservoir for surplus vitality. Others believe that it is engaged in the manufacturing of white blood corpuscles. Whatever may be its office, we know that mental disturbances and powerful drafts upon the nervous energies have an unfavorable influence upon it, giving rise to serious disturbances in its locality, which are usually construed into heartaches or heart difficulty. It is therefore well to understand the relative position of the two organs in order to determine which is affected.

It must be remembered that the heart is two or three inches higher than the spleen and more nearly in the centre of the chest. All suspense, anxiety, and worriment of mind brings the distressed leaden weight in the left side below the breast. The distress and its locality will be readily recognized by all that have suffered from mental anxiety. This oppression and distress doubtless comes from an engorgement of the splenic blood vessels, it being in close sympathy with the brain from its intimate connection with the great pneumo-gastric plexus.

We speak more fully on this subject, in order to relieve anxiety concerning the heart difficulty, for although people sometimes become frightfully disagreeable through splenic difficulty, they rarely die of it, whereas in heart diseases they occasionally do.

According to Bosquillon, the spleen is engaged in generating what is known as animal magnetism, those persons being most magnetic in which that organ is largest and healthiest. As the world of mankind is largely controlled by the magnetic power, if this man's theory be true, it would be well to increase the capacity of that portion of the body in which that organ resides, to give it full play in the performance of its functions.

TREATMENT OF PILES.

In all cases the aggravating cause must be removed, be it weight of clothing, constipation, long continued standing upon the feet, running up and down stairs, etc., as all remedies must fail if these wrongs are continued. The most effectual remedies are enemas of tar water, or the oil of tar applied immediately to the parts. In hemorrhoidal piles, cotton-wool may be saturated with oil of tar and applied; hot injections should be used frequently to allay the irritation, but where there is great weakness in the mucous lining a salve composed of mutton tallow, rosin, sweet clover blossoms, and a small quantity of tannin may be made into pastiles and introduced into the rectum upon retiring.

TREATMENT OF BLADDER AND KIDNEY DIFFICULTIES.

An irritation of the mucous lining of the bladder and its mouth is most frequently caused by some form of kidney difficulty; still, prolapsus of the bowels or uterus will often produce all the aggravated symptoms of that disease, by the heated portions pressing against the walls of the bladder, preventing a proper secretion of the urine, causing a desire to urinate frequently,

45

giving rise to severe irritation **of the** neck and mouth **of** the bladder, **which at times** can **only** be allayed **by** injecting into this organ some cleansing **and** soothing lotion.   **This can be readily** done **by a** small ear syringe.   **The** mouth of the bladder **can be** easily **found** as it opens immediately above the vagina, and the delicate pipe can **be** introduced without giving pain. This should always be resorted **to** before the **inflamma-** tion becomes deep-seated.   **Quite a** strong solution **of** borax **and** water will be found **an** excellent lotion.

To keep **the kidneys** in health that portion of **the back** where they are located should be daily rubbed **and** manipulated; **the** same treatment we **have** advised for the **liver.   The kidneys do not lie as** low **as is** generally supposed.   **By turning the hands** backward **and** clasp- ing the **waist at the back, the fingers** press directly over them.   **With a steady,** firm pressure rub inward **toward the** spinal **column, the** direction in which the tubules which carry the urine to the bladder pass. This treatment tends to dislodge any limy sediment that may collect in the delicate ducts, and is especially **necessary for those** who **do not take** much physical exercise **as the** ordinary healthful movement of the muscles **will** perform this work naturally.

Where **the** kidneys **have become** weakened or dis- eased, mineral waters should be resorted to, **and** in case of granulation the use of koumiss, or sour milk, butter-milk, cottage cheese, etc., as the lactic acid contained in these articles has the power to disintegrate limy col-lections in the system.

In case of diabetes, daily hip baths should be taken, and at night wet compresses enveloped in dry flannel should be worn over the loins.